HEALTH AND EXERCISE IS WEALTH WITH "RECIPES"

SALLY IYOBEBE

Trafford rev. 02/05/2016

 www.trafford.com

North America & international
toll-free: 1 888 232 4444 (USA & Canada)
fax: 812 355 4082

DEDICATION

THIS BOOK IS DEDICATED TO THE ALMIGHTY GOD FOR
THE STRENGHT OF HARD WORK AND DEDICATION WHICH
MADE IT A REALITY.

ACKNOWLEDGEMENT

FIRST AND FOREMOST MY GRATITUDE WILL ALWAYS BE THAT OF THE ALMIGHTY GOD WITHOUT WISH IT WON'T BE POSSIBLE TO WRITE THIS BOOK.

SPECIAL THANKS TO MY HUSBAND, MR. KINGSLEY IYOBEBE (THE KING) FOR REVIEWING SOME OF THE MATERIALS USED IN THISBOOK.

THANKS TO MY SISTER MAUREEN AFUE AND MY DAUGHTER MARGARET FAGBEMI FOR HELPING ME WITH SOME OF THE RECIPE USED IN THE BOOK AND PRINCELY BESONG FOR BEING THERE EACH TIME I NEEDED ASSISTANCE IN THE COURSE OF WRITING THIS BOOK.

SPECIAL THANKS TO OUR FAMILY DOCTOR, DR. OLUYINKA ADEDIJI, MY AUNTY AND PASTOR, PASTOR KEMI SEARCY AND MY BOSSOM FRIEND BARRISTER ONII NWANGWU-STEVENSON FOR AGREEING TO USE THEIR PRECIOUS TIME TO WRITE THE PREFACE. (FOREWORD) OF THIS BOOK.

MY SPECIAL ACKNOWLEDGEMENT TO MS. KAREN FRAZIER FOR USING HER TIME TO PROOFREAD THE BOOK. AND FOR LOVETT IYOBEBE WHO HELPED IN TYPING THE SCRIPTS. THANK YOU GUYS SO MUCH FOR THE WONDERFUL JOB.

MANY, MANY THANKS TO MY SWEET MOTHER, MRS. ANNA BESONG AFUE, A RETIRED SCHOOL TEACHER AND A STRONG WOMAN OF GOD FOR TEACHING ME NOT JUST TO LOOK GOOD AS A WOMAN, BUT ALSO WHAT IT

TAKES TO BE A GOOD HOUSE WIFE AND A GOOD COOK. SHE ALSO IMBIBE IN ME THE ATTITUDE OF SCHOOLING, PRAYING, CLEANING, FARMING, AND ABOVE ALL BEING INDUSTRIOUS.

FINALLY TO LATE PA JEROME TILLEY –GYADO, CHAIRMAN OF GYADO GROUP OF COMPANIES, MY GOD FATHER. HE GAVE ME MY FIRST JOB AS A RADIO OPERATOR AND SPONSORED ME FOR 4 YEARS AT THE COLLEGE OF AGRICULTURE WHERE I EARNED THE OND AND HND IN HOME ECONOMICS. BECAME THE YOUNGEST MANAGER IN HIS HOTEL AS THE CATERING AND RESTAURANT SUPERVISOR. I WILL ALWAYS BE GRATEFUL TO HIM. MAY HIS SOUL REST IN PERFECT PEACE AMEN!

INTRODUCTION

When I wrote my first book on "Healthy Eating and Lifestyle", it was all about trying to lose weight, but a lot of readers suggested that I should write about the food (diet) and exercises that are necessary to achieve "Healthy Eating and Lifestyle". So I decide to write this book in answer to those reader's request.

I decided to bring into play my vast experience as a graduate of Home Economics from the College of Agriculture, Yandev-Gboko, Benue-State, Nigeria. An Owner/CEO Caesar's Palace Restaurant, Gboko, Benue-State, Nigeria. And a teacher of Food and Nutrition at the Government Girls College, Mabera, Sokoto State, Nigeria.

The first time I decided to lose weight I joined, the Virgin Miles Club and started exercising seriously until I was able to cut down 70 pounds in six months. This was done through determination, by eating right, exercises, and fasting and prayers. I grow my own organic vegetables, cooked my own food.

I understand a lot of people hate cooking because of the many trouble in it. You do not have to cook every day. Cook enough that you can store for a couple of days in your refrigerator and heat it up whenever you want to eat or hungry. Try to discourage yourself from eating all the ready-made food in the cans because you do not know how it was prepared and in what type of environment. As you know, nothing good comes easy. You can go for an entire hamburger, it is easy to add so much weight but at the end of the day, you pay so much to lose it.

This book is full of different recipes mostly from Africa, is worthwhile trying it because you will never regret it. I will be willing to put you through on any of the recipes you may want to try your hands on. And will also help locate where you can buy the products.

In the area of exercises, I will be willing to exercise with you, only if you are ready and willing to help yourself. Let me know how you want me to help. I will be willing to help. Check the different recipes in the book.

While I acknowledge those who are out there trying to sell the "get slim" capsule or pills, I will urge you to do it naturally by exercising, eating, cooking your own food and making good life choices. Fast once in a while and you will definitely see a better you.

Sally Iyobebe, HND, LL. B, MBA, MSLS

Thanks, God Bless and More Grace.

Sally Iyobebe.

President, Princess of Zion Center, Inc. Montgomery, Alabama USA and Princess of Zion Charitable Foundation, Lagos Nigeria

To those of you who want to learn how to cook, arrangement can be done for a fee.

If you want the food to be prepared and brought to your house, it can also be done for a fee.

For graduation, wedding, birthday, or any occasion contact me.

God Bless and More Grace.

Sally Iyobebe

FOREWORD

In this concise book, Sally has eloquently laid out the fundamental factors leading to poor health and proffers means to solving them. Health and Exercise is Wealth is a master piece of work from an author who is blessed with a persuasive oratory. She began by properly settling the true definition of health as it is simply understood to the comprehensive definition from the World Health organization - WHO. This fundamental approach by itself, figuratively underscores the enormity of the health crisis we face today.

The author understands that in order to make a permanent inroad in the universal health crisis, we have to succeed with the individual's one at a time and by so doing, the family, the community and the whole nation in general.

In this book, we have all that is needed to sensitize us in terms of making the case for health with emphasis on Obesity, Diabetes Mellitus, Stroke, Cancer and Arthritis - chronic diseases which if not controlled, can bankrupt the United States of America. She went on to enjoin us to embrace exercise while we choose what we eat economically and wisely. The chapters on diabetes, exercise and the effect of certain food in fat genesis are simply worded without the nerve-wrecking details of a scientific publication. This style of gentle persuasion is a unique talent of the author. She brought to life her love of science with a blend of culture and the preeminence of women involvement in our battle against the obesity related diseases ravaging American and entire world families.

As we progress through this book, we are introduced to a set of concisely described activities that benefit the body immensely, including the intensely physical ones like jogging to the quiet but mentally intense practice of yoga. Sally challenges us to reject the notion of "wear and tear syndrome" but instead embrace the uplifting concept of "use it or lose it." When it comes to exercise, the author believes that - more is better. The author's spirituality was evidently revealed in this book. The notion of self-restraint and self-denial by our biblical ancestors even if for many different reasons resulted in better health which therefore serving as template to many ideas of dieting and fasting that are being observed today.

The third and final aspect of this book focuses on healthy food items including preparation instructions. This approach of step by step guide makes this book more valuable than any health guide on the market today. It simply means that one can start out having thoroughly understood the role played by exercise and food choices, then select a menu, follow the instructions and you have on your dinner table a meal that is loaded with vital and spiritual energy.

The all in one cook book which features several time tested African menu is a valuable asset to individuals, families and groups that seek a lasting solution to our collective health problems.

After accepting to write this introduction, I began to question my qualifications to introduce the work of our highly esteemed sister, wife, mother, educator, and counselor. Of course, it is too late for me. I must confess that this book has opened my eyes and greatly enriched my understanding of the interconnection among exercise, spirituality, healthy eating and health as it is meant to be by our creator.

I recommend this great book to all and I believe it would change your life for the better.

Oluyinka Adediji - a physician, published poet and health disparity champion.

FOREWORD

Are you ready? You are about to embark on a journey that can change your life. "Good Health and exercise is Wealth" is a "Treasure" A Pearl of Great Price. I am sure you are aware that wealth without health amounts to nothing. It amazes me when I consider our nation with all of its technological and medical advancements, and yet we still remain one of the unhealthiest nations on the earth. I can't help but to agree with the saying "You are what you eat." You really are what you eat. The majority of the ills of our society, as far as health is concerned, can be attributed to our dietary lifestyles. Poor Diet and lack of exercise contribute to illnesses such as High blood Pressure, Sugar Diabetes, High Cholesterol, Strokes, Mood Swings, etc.

It is time to take control of our lives, by taking charge of what we put in our bodies. We must respect the body as the temple of the Most High, and treat it as such. In this book Ms. Iyobebe has beautifully outlined God's road map for a healthier you. She highlights the right grains, meats, fruits and vegetables, and the right way to prepare continental dishes that will produce the right nutrients the body actually needs to function at its maximum productivity. This is a must read for all, especially those of us with a call of God on our lives. No matter how anointed you are if your body is broken, and out of shape there is very little you can do for yourself and the kingdom of God. Don't forget you carry God's treasure in earthen vessels. Follow the principles laid out systematically to a healthier and a prettier you.

PASTOR KEMI SEARCY
Fresh Anointing HOUSE OF WORSHIP
150 E. FLEMING RD,
MONTGOMERY, AL 36105.

FOREWORD

I recently asked my son if he had a good time at his friend's birthday bash, he replied in the affirmative, but said his legs were aching from dancing. He also said his friends said he had lost his dancing skills. I told him the reason was obvious – he had become overweight. Then I read this beautiful food-for-thought book, and I could see why it is a must read for him and anyone desirous of a healthy lifestyle.

Nine chapters of detailed expose into healthy living and healthy lifestyle. It is health management at a glance as everything you need or require to attain and sustain good health is laid bare in very simple and easily comprehensible prose that captivates the reader's attention.

The author shows a wide and versatile understanding of the essentials nature has provided to aid our quest and desire for a healthy lifestyle.

A detailed guide on how to make good choices to ensure good health, if you ever felt concerned about your weight, this book is an inspirational guide on how you can prevent the ugly consequences by making the healthy lifestyle choices. The content is both theoretical and practical.

The major strength of the book is in its use of diverse universal materials and examples to illustrate situations. The portion on recipes is also varied and comprehensive with wide range of varieties to select from. The book is therefore appropriate and more relevant to our needs compared to similar works that rely on examples and materials that are confined within limited social and cultural realities and experiences.

The author in this book exhibited clear understanding of the issues and challenges associated with the efforts to attain and maintain healthy lifestyle. Her years of experience in health care and nutrition, plus her sojourn in Africa and the USA have a definite influence on the style and dexterity with which she presented each chapter in the book.

Beyond theory, the author spent ample time and space in practical demonstration of steps to healthy life style.

This book should be part of basic readings for healthy management studies in schools; and a handbook for anyone who desires to eat his/her "way to excellent life choices".

Onii Nwangwu-Stevenson (Barrister at law)
Principal Partner,
Options and Equity Chambers
Lagos – Nigeria

CHAPTER ONE

WHAT IS HEALTH?

To every individual, health has so many definition and have several meanings depending on who you ask. A layman might define health as when one has total control of their well-being. Health can be physical, mental, and the ability to be free from any disease that could kill. Physical health could be any activity that might affect any part of the body and prevent it from performing its function. While mental health affects the ability to think and be of sound mind.

Apart from medical definition, the health and well-being of a country can be measure economically by reviewing their Gross Domestic Product (GDP). A country full of new and old business prospective, has a healthy economic that encourages investors as well as increase employment. Just as health is important to the economics of a country, it should play a vital role in our life both consciously and unconsciously.

Over half a century ago, the Word Health Organization (WHO) defined health as "a complete state of physical, mental, and social well-being, and not merely the absence of disease or infirmity" The quote is commonly used as a definition of health.

To help you lose weight and keep it off, you have to gain more knowledge about how your body works. In a human body, there are muscles and bone which assists in the balancing of the body framework for steadfastness. We have to listen to our body in order to predict a condition that might cause harm to the body. Listening to our body would help us plan a defensive attack against any sickness.

According to popular saying, "health is wealth". This mean we should watch what we eat and ingest into our body. In essence "we are what we eat". We should only eat food that will benefit our health. There is another saying that we should always have at the back of our mind, and it says "one man's meat is another man poison". So do not eat everything you see other people are eating, and because it looks good does not mean it is good for you. Do not forget, if you want to lose weight and keep it off, you have to exercise, and change the way you see food. Also pay attention to the amount of calories, food size, and fat necessary in helping to transform your body to a better you. It is also good to eat food that are rich in fiber, which will help build your body in form of calcium, fiber, and carotene. Remember, too much of everything is bad, so moderation is the key to a healthy lifestyle. Eating healthy combine with exercises can help with quick recovering from many sicknesses like body pain, diabetes, arthritis etc. Below is the list of food that are rich in fiber:

FRUITS: Prunes, pears, mango, apples, raspberries, raw blackberries, raw strawberries, seedless raisins

BREAD: Rye, Bran flakes, wheat bread.

NUT & SEED: Almonds, pistachios, peanuts, pecans.

LEGUMES: Navy beans, pinto beans, black beans, lentils, kidney beans, lima beans, split peas.

Being hungry and having appetite for a particular food are two different things that are used interchangeably. When you eat from wanting a particular food, instead of eating from being hungry is called temptation that pushed people eat and restore calorie. These extra calories increase the waistline and dress sizes.

This is where discipline comes in, because Self-discipline is very important when the need to eat for pleasure arises. Self-discipline is not only in the area of eating too much, for we must also discipline our body in other way, as our body is God's temple. Constant exercise is good for the body, just as for lack of exercise will make the movements of the muscle and joints stiff, especially when one is getting older, and the body is no longer as flexible as it used to be.

One can do exercises by doing chores, picking weeds from the garden, walk at the mall and in the park. For some they can go to the gym, walk out early in the morning and in the evening for about 30 minutes a day to help calm down the mind, emotion and thought process. This is because our health is geared towards our mindsets. Therefore, we strive to think positively at all time to free our minds and body from emotional stress, anxieties and tensions.

While it is advisable to eat fruits as part of weight lost, too much of it can increase blood sugar and makes you very uncomfortable as the different is not defined in the body between the fructose and natural sugar. So eat less of fruits. Your eating pattern needs to be changed or altered for a long period. This is because you need to concentrate on the kind of food you eat and the lifestyle you want to maintain. And this should be mainly on the part to a healthy food choices, desire it all the time and in turn you will be amazed how you just lose weight without having to struggle. Make sure you eat a lot of vegetables as part of the menu for a healthy living. And drink a lot of water to help full your stomach and make sure there is no space for any food.

CHAPTER TWO

KEY TO GOOD HEALTH LIES IN MODERATION AND BALANCE

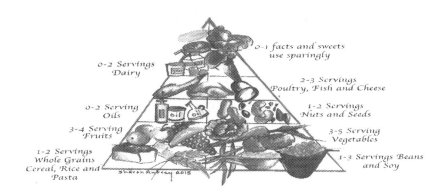

What you eat affects you is true to the old adage "you are what you eat." This is true as healthy eating is the corner stone to a healthy wellbeing or life. Going back to Chapter four of the healthy eating and Lifestyle, the body is made up of about 100 trillion cells of which each demands a constant supply of daily nutrients in order to function well. Food affects all of these cells and by extension, every aspect of our wellbeing, mood, energy levels, food cravings, thinking capacity, sex drive, sleeping habits and general health. When the body is fed with all kinds of junk and convenience foods, it will definitely lay down fat, lower your energy and in most cases diminish your brain power.

Though the center for nutrition, Policy and Promotion have changed MyPyramid and developed the "My Plate" model to help remind us to eat healthier. The "My Plate model was designed to remind everyone to eat healthfully. This is not meant to alter any one's eating pattern or habit. It just focuses on the five food groups necessary to help individual make some better food choices for a healthy living. Credits should be given to Mrs. Michelle Obama, (First lady of the United State), for her effort in implementing my plate and trying to reduce the rate of obesity in young adult and children. She also encourages everyone in America to make a healthy food choices and the right food in their plate during meal time. Though in this book, I will be using my pyramid as part of the illustrations of the five food groups for a balance diet. The design of MyPyramid consists of vertical colored stripes. Each color has a different size, suggesting the amount of food that you should choose from each group. The figure on the stairs is there to remind you of the importance of physical activity.

MYPYRAMID-GRAINS

In my Pyramid grains, it is suggested that one should eat at most fifty to sixty percent of the whole grain, made up of whole grain bread, cereals, crackers, pasta or rice. Grains are of two types: The refined grain and the whole grain. One is advised to eat more of whole grain because the refined grains which is milted for finer texture, takes away all the fiber, iron and vitamin B that helps in the reduction of cholesterol and of the digestive system.

MYPYRAMID-VEGETABLES

In MyPyramid vegetables, it is suggested that, the dark green vegetables, dry beans and peas should be eating more. The amount of vegetables one eat should be recommended by their physicians, or it varies with their respective ages, sexual and physical level of activities. Vegetables is the best source of vitamin B and C and it should be eaten every day. Vegetables can be blended and drink as juice or smoothie.

MYPYRAMID-FRUITS

In My Pyramid fruits, it is suggested that one eat different kind of fruits and drink less of fruit juice. Fruits could be eating fresh, frozen, canned or dried. However, just like vegetables, the quantity of fruits eaten should varies with one's age, sex and physical level. One is advised to eat at most two to three fruits a day as it contains several vitamins and minerals that is necessary for the nourishment of the body and a good source of fiber. Fresh fruits are better and healthier to eat than drinking, because it contains a lot of sugar. You can also make your own juice from the fresh fruits as it is natural without adding any sugar in it. It will serve the same purpose as eating fresh fruits. A glass of fresh juice in the morning before going to work or school is recommended in the mornings.

MYPYRAMID-OILS

Oils from plant like vegetables and nuts are good for health as they do not contain cholesterol. Butter and margarine should be avoided if one can as they come from animals and contains too much fat which contributes to cardiovascular disease. This disease is one of the leading cause of death in the USA. Most fat eaten should come from vegetables, nuts and fish.

MYPYRAMID-Milk, Yogurt and Cheese

In MyPyramid-milk, yogurt and cheese, it is advisable to drink fat free or low fat milk. Milk, yogurt and cheese are all good for the body as they are rich in calcium. Drink lactose- free product and other calcium products as it is necessary for bone growth. But cream cheese, cream, sweetened milk and butter should be avoided as they contain extra amount of calories. If you cannot drink milk, you can always drink or eat other food products that contain calcium.

MyPyramid recommends 3 cups per day of fat-free or low-fat milk, or milk products for adults.

Children 2-8 years old should consume 2 cups per day of fat-free or low-fat milk.

Milk, yogurt and cheese are all rich in calcium.

Choose lactose-free product or other calcium sources if you can't consume milk.

Calcium is important for developing bones, especially when you are an adult. Milk, yogurt and cheese are all rich in calcium. Be aware though that cream cheese, cream and butter are not rich in calcium. You should also be aware of the extra calories that contains in the sweetened milk products that you choose. If you can't drink milk, try lactose free products or other calcium sources.

MYPYRAMID-Meat, Poultry, Fish, Dry Beans, Eggs and Nuts

In My Pyramid meat, poultry, fish, dry beans, eggs and nuts, it is advisable to eat low fat or lean meat and poultry. Avoid any fried meat or chicken. You should rather roast, bake, broil, or grill your meat, chicken or hen or fish. Fish, beans, nuts peas, and seeds of any kind should be eating in large quantity. While some people may like to eat red meat, they should alternate it with fish, beans, peas, nuts and seed as they contain or have a good, healthier unsaturated fats. Salmon, trout and herring are good for the body as they are high in Omega-3 fatty acids, which are good for the body as well as good health. Flax and walnuts are good sources of important fatty acids. Sun flower seeds, almonds, hazelnuts are good in vitamin E.

MYPYRAMID– Physical Activity

In this MyPyramid of physical activity, it is important to go for a walk in the park or around your neighborhood 30 minutes a day. At your place of work, you can use your break time and go for a 30 minutes' walk. This will help in given you a healthy body and make you physically fit in carrying out the day's activities. Children and teenagers should also be advice to carry out physical activities so that they can be physically active both at home and in school. Families should be advise to carry out physical activities as a group as to encourage one another.

The plate

From the above food MyPyramid, to form good nutritional habits, you have to track what is in your plate. Learn to make good food choices as good nutrition plus physical activities are the essential ingredients to a healthy lifestyle. A nice plate should be full of all colors of food. The plate should have more vegetables and meat as big as a size of the deck of card. Obesity rates among all groups in society, irrespective of age, sex, race, ethnicity, socioeconomic status, educational level or geographical region, have increased remarkably.

What causes Overweight and Obesity?

Weight gains depends on the amount of calories intakes, the amounts stored in the body and the quantity that is burnt up during exercises. However, overweight or obesity can be influenced either through genes or environments. They can both have effects on one's physiological and behavioral pattern. Some people might decide to eat pizza, hamburger, cakes and still be thin. While some people like us will eat the same food and even in smaller quantity and still add weight or grow fat.

Environment

Environment plays as a biggest contributory factors to weight gain and obese. This is because people eat too much and never had the time to exercise and when they do, it is too little. It will be said that, where people resides, go to school, work and play, have a lot of impact on people being active. People are so used to driving to the store even when the distance from their houses to the store is just about five to ten minutes' walk from where they reside. It is very necessary if one has a bike, use it to work, or at a building with a stair, use the stairs instead of the lift, at the grocery store, park far off so that you can walk to the store. At your place of work, during your break time, use 15 minutes to 30minutes in walking. Early in the morning, go to the park with friends, families or co-workers and walk. Even within your neighborhoods, the streets are build's in such a way that makes it easily for one to walk. Where one is physically active, it adds a lot of numerous benefits to one's health.

When a woman is pregnant, and smoke, the baby is likely to be overweight. Drinking of too much sodas and the eating of high calories, processed foods are on the path of becoming overweight. Too much of static activities like watching of TV and video game are a pattern for becoming over weight.

In 2012, some of us from my Church, Fresh Anointing House of Worship in Montgomery, Alabama, including my Pastor, Pastor Kemi Searcy and Bishop Fred Adejunji, travelled to Israel for a mission trip. During those two weeks of that trip, we walked and climb mountains, without even feeling it. something we will never do while in the USA. We never envisaged that kind of a walk, but at the same time, we really enjoyed it as it helps to burn a lot of calories from our body. Instead of looking rough, we were looking healthy just like Daniel did when he ate only vegetables and drank water for ten days without the king's diet and wine.

Genetics and Obesity

It is true that genes contribute to a person's developmental stage. At the same time, it is difficult to pinpoint to that genes specifically. However, eating of fried foods can help a person's genes in contributing to obesity but only in a small amount. But when a person lives a healthy lifestyle it can destroy everything about genetics. As this affect all classes of people whether you are poor or rich, or illiterate and educated. This is because people physics defers from one another and depending on how you want to live your life, you can always control your body fat.

Also depending on the kind of illness or drugs one takes, there is tendencies to add to one's weight. It is advisable to see your doctor as to what is the contributing factors to your gaining weight.

Sally Iyobebe

What are the consequences of overweight and obesity?

There are a lot of consequences as it relates to the disease of the heart and might cause stroke. Because it is left to go out of control, it causes a lot of suffering to that person and pains to the families as well. There is a financial implications relating to the costs of health care which in many cases is on the increase. Apart from health costs, even where you are force to start eating healthy, there is still the financial implications to balance diet relating to eating the right stuff. Vegetables, fruits, legumes, whole grains and nuts are quiet a good preventive measures of overweight. Avoid the intake of sugars as it is not good for the body. It is always good to check the amount of calories intake without forgotten to do at least 30 minutes of either regular, moderate or intense exercises or activities every day as to chop off extra fat in the body. Excess of fat may lead to type 2 diabetes, high blood pressure and the added high cholesterol in the body. It may also result to cancers which could be in the form of breast, colon, or endometrial. It is better to quickly addressed these issues before it goes out of control.

How to Prevent overweight

Being overweight increases the risk of type 2 diabetes, heart disease, high blood pressure, stroke, high cholesterol and high blood glucose. Being adhering to a healthy eating formula by healthy lifestyle choices. One needs to be active by doing house work, go to the park for a walk or do some running. One should spend less time on computers except if you are using it for work purposes or for school, TVs and keep track of the amount of weight loss and the body mass index.

Build a Healthier Plate

In order to avoid the risk of type 2 diabetes and heart disease, it is advisable for one to eat the right meal so as to maintain a regular healthy weight. Some families or people have monetary shortfall to stay on a healthy diet and live a healthy life style. They can try making a shopping list of the kind of food stuffs they want to buy before going to the store.

And in that list, there should be more of fresh vegetables and fruits. There should be fresh fish, lean meats like turkey, chicken, lean cuts of beef, low fat dairy products like skim milk, yogurts and whole grain bread and cereals. Avoid the drinking of soda, sweet drinks, French fries, chips and other snacks food that are fattening.

Shop Smart.

Smart shopping for meals is very important as it requires you to think and make wise decision before going for grocery shopping. Therefore, you might want to create time by taking inventory of all that you want to buy before going for grocery shopping. This can be done either weekly, bi weekly, monthly to save you a lot of money, time, and stress as well as avoid too many trip to the store. You should try as much as possible to avoid any canned or frozen food as it is not healthy. And if you must buy canned vegetable or fruits, make sure you drain all the water and rinse it with fresh water. Some other food stuffs to buy as part of healthy diet are brown rice, whole grain foods, crackers (unsalted) and less sugar, and cereals. When shopping never be tempted to go outside the grocery list

Eat Smart.

Eat plenty of green salad and tomatoes, broth soup with plenty of vegetables which can fill your stomach. One is recommended to eat a lot of grilled meat, fish, chicken, turkey and avoid any kind of fried food. Avoid foods that contain high calories and those with low vitamin and minerals. Avoid any canned vegetables, juices, and rather buy fresh fruits and vegetables and make your own juice which is natural without any sugar added to it. But again avoid eating too much fruits because of the sugary content. The body does not know the difference between natural sugar and fructose sugar. And drink plenty of water to help flush out any fatty or sugary substances in the body. You can also prepare your own home made snacks with plenty of vegetables added to it to make it a healthy meal. When you eat out with your family or friends, try and skip dessert. There is nothing like a little will not harm you or me. It will definitely harm you. It is just like you have $1.000.000. When you take away 10 cents from that amount, what happened? It has already harm the $1.000.000 because it is no longer a round figure. So also the little stuff we eat and take for granted.

Small Steps.

Before one start taken any small steps at all, you must first of all consult with your doctor for any advice on weight lost. To starts weight exercises, one has to first take some small basic steps as against longer or larger steps, rather than trying to do it all at one time. One must have to set an objective that are achievable within a reasonable period that one thinks it will be convenient for him or her. As one does it, make sure one has a planner to keep track of all the necessary records of the exercise and the amount of weight loss one want to achieve. Have a different note book to write down all the food and liquid intake, which has to include the portions necessary to eat. One might also want to keep track of one's daily portion of calories or fat that one eats. It is also wise while in the store for grocery shopping, to look at the amount of calories on the label. Where it is more than 140 calories, it is advisable that one try to avoid it. And if one must take or eat it, one might have to do some exercises to break down those calories or fats. It is recommended that one check one's weight once a week, write it in his or her planner and make sure one buy smaller size of clothes as this will help one in reducing one's weight quickly.

If one is the lazy type that cannot do exercise by one's self, then get a group or form a club that can motivate you. Make sure they too in that group are also ready to lose weight. Buy exercise tapes, play them and follow the steps. Bigger loser tape is recommended if you are really motivated to lose weight quickly. At one's job, one can form a support group of people having weight problem and who are willing to lose weight and together go for a walk during their break time and they will be amazed on how effective it will be. I encourage everyone, not to look back, so that one can actualized one's lose weight goals or objectives.

CHAPTER THREE

DIABETES

Diabetes is a disease that is commonly found largely in our society and affects the way the body uses food for energy. Normally, the sugar you take in is digested and broken down to a simple sugar, known as glucose. The glucose then circulates in your blood where it waits to enter cells to be used as fuel. Insulin, a hormone produced by the pancreas, helps move the glucose into cells. A healthy pancreas adjusts the amount of insulin based on the level of glucose. But, if you have diabetes, this process breaks down, and blood sugar levels become too high.

There are two main types of full-blown diabetes. People with Type 1 diabetes are completely unable to produce insulin. People with Type 2 diabetes can produce insulin, but their cells don't respond to it. In either case, the glucose can't move into the cells and blood glucose levels can become high. Over time, these high glucose levels can cause serious complications.

Pre-Diabetes:

Pre-diabetes means that the cells in your body are becoming resistant to insulin or your pancreas is not producing as much insulin as required. Your blood glucose levels are higher than normal, but not high enough to be called diabetes. This is also known as "impaired fasting glucose" or "impaired glucose tolerance". A diagnosis of pre-diabetes is a warning sign that diabetes will develop later. The good news: You can prevent the development of Type 2 diabetes by losing weight, making changes in your diet and exercising.

Type 1 Diabetes:

A person with Type 1 diabetes cannot make any insulin. Type 1 most often occurs before age 30, but may strike at any age. Type 1 can be caused by a genetic disorder. The origins of Type 1 are not fully understood, and there are several theories. But all of the possible causes still have the same end result: The pancreas produces very little or no insulin anymore. Frequent insulin injections are needed for Type 1.

Type 2 Diabetes:

A person with Type 2 diabetes has adequate insulin, but the cells have become resistant to it. Type 2 usually occurs in adults over 35 years old, but can affect anyone, including children. The National Institutes of state that 95 percent of all diabetes cases are Type 2. Why? It's a lifestyle disease, triggered by obesity, a lack of exercise, increased age and to some degree, genetic predisposition.

Gestational Diabetes:

Gestational diabetes (GD) affects about 4 percent of all pregnant women. It usually appears during the second trimester and disappears after the birth of the baby.

Like Type 1 and Type 2, your body can't use glucose effectively and blood glucose levels get too high. When GD is not controlled, complications can affect both mother and the baby. Your doctor will help you work out a diet and exercise plan, and possibly medication. Having GD increases your risk for developing it again during future pregnancies and also raises your risk of Type 2 diabetes later in life.

Natural Remedy for curing diabetes.

There are so many ways of curing diabetes naturally. This is necessary when one has the believe going through the natural remedy will help in curing the diabetes. One of the best way of curing diabetes is for one to eat a balance diets and avoid the eating of too much of junk and sugary food as it will help in the increase of the blood sugar. Added to that, doing exercises at least 30 minutes every day will help in curing the diabetes. While it is true that overweight can give birth to diabetes, there are some percentage of people who are suffering from diabetes and are not over weight. This is as a result of heredity as being sickness in the family.

Another way of curing diabetes naturally is by eating plenty of vegetables diets, organic fruits and not the ones in the container that has a lot of sugar in it, and drinking of plenty of water to flush out excess sugar from the body. Also whole grain foods, avocados, beans, sweet potatoes and soy milk are of great help in curbing diabetes. Cinnamon which serve as spice in foods has added flavor by eating it at least twice a day can also help in the reduction of sugar in the blood.

Herbal Remedies

There is an array of herbal treatments that can be used as natural cures for diabetes. Taken as daily supplements, herbs promote circulation, glucose maintenance, and cell growth in diabetics. The Indian herb, Gymnema Sylvester, lowers blood-sugar levels and reduces sugar cravings by stimulating the pancreas to naturally produce more insulin. Bluberry leaf extracts work to rejuvenate damaged cells and stimulate the growth of healthy cells, reducing the incidence of macular degeneration. Cayenne is known to increase blood circulation so that oxygen can reach blood vessels and tissues, decreasing the risks of infection, retinopathy, and limb amputation. Natural herbal cures offer significant support for diabetics.

Daily Exercise

Daily exercise leads to healthy maintenance of the body's systems and organs. Regular fitness regimens promote healthy weight, increased circulation, and stress relief. Weight training and stretching exercises elevate joint flexibility, blood flow to the limbs and extremities, and muscle building. Aerobic activities such as running, walking, and swimming help metabolize stored glucose in the bloodstream by converting it to energy. This process results in normal sugar levels, improved insulin response, and healthy weight.

Endorphins released through exercise work to strengthen the immune system while combating diabetes symptoms of fatigue, anxiety, and depression that can arise from hypoglycemic episodes and insulin side effects. Exercise combined with a nutritious diet can be used effectively to control and prevent Diabetes Type 2.

Diet and Monitoring

A healthy diet for a diabetic will increase metabolism while monitoring the amount of sugar that enters the body. Foods rich in carbohydrates such as; bread, pasta, cereal, yogurt, and fruit spike sugar levels immediately and send more sugar into the bloodstream. Dependent upon the amount of medication, serving size, and pre-meal sugar levels, diabetics will respond to differently to carbohydrates at various times of the day. Carbohydrate intake should be carefully allotted and planned in accordance to blood sugar levels and dietician recommendations.

A healthy meal plan for a diabetic should be high in fiber and low in fat, as fiber-rich diets allow for less sugar absorption and low-fat diets control weight. Whole grains, fruits, vegetables, beans, brown rice, lean meat and low-fat dairy products are healthy dietary choices. Fruit and plain, low-fat yogurt are nutritious sources of fiber, minerals and vitamins. Both carbohydrates offer healthy snack choices if consumed in controlled portions after glucose levels have been monitored and are within the low to normal range.

Blood sugar should be monitored regularly upon waking up, before each meal, an hour after each meal, and prior to bedtime. Readings are important in maintaining healthy sugar levels, and assist in meal planning and exercise. Fasting and pre-meal readings in the range of 70-to-110 are normal. One hour after a meal, diabetics should aim to reach a reading of 140 or less. Monitoring is a natural cure that provides diabetics with a concrete understanding of how food effects sugar, so that nutritious and beneficial meal plans that target healthy levels can be implemented.

Steps in curing Diabetes

Too much of fast food leads to diabetes, and America is now embedded in both. According to a new report by the CDC, that by 2050, nearly one-third of American adults will be diabetic. This has become a windfall of profits for the drug companies, diabetic supply companies and the sick care industry in general. Perhaps that is one of the reasons why no one is taking any real action to halt the anticipated explosion in diabetes. Except perhaps for the First Lady of America, Mrs. Michelle Obama who has devoted most of her time since coming into the white house in making sure that people eat right and exercise in cutting down their weight and to prevent diabetes as well. The more people who get sick, after all, the more money will be spent on medical treatment. And this is good for the economy.

Causes of Diabetes

Heredity:

1. Heredity plays a role in the spread of diabetes while it has not been scientifically proven how it sometimes runs in the family. However, not every one that is born in that family will have the disease and even one not born into the family of diabetes can equally develop one. Though those born into that family of diabetes may have the chances of acquiring the disease, but can be prevented if they adhere to a proper preventive measures.

2. Wrong Mediations- Because hospital bills are on the increase, a lot of people will rather go buy drugs at the drug stores rather than see a doctor. They end up using the wrong medication which can damage their pancreas. Drugs generally buying from above the counter are dangerous and should be taken with proper precautions. Always seek the advice of your doctor before buying any drugs above the counter.

3. Lifestyle: Wrong lifestyles are some of the major causes of the disease to a greater extent. But people do not want to listen. As food can nourish one's life, it can equally destroy it. Some mothers are prone to feeding their children with too much of soda drinks, sugar, ice creams, biscuits, chocolates and refined food. They may think it is part of civilization, but indirectly they are destroying the life of the child/or children. Drinking of beer, expensive wines and eating of too much red meat, drinking of coffee and adding too much sugar in it, signifies that the person is eating himself or herself to the grave.

4. Overweight: When a person takes too much of proteins food and heavy caloric food, it will help the person gain weight which will lead to overweight.

Symptoms of Diabetes:

One should try and do whatever they can to prevent diabetes. Here are some symptoms of diabetes:

1. Dryness of mouth and excessive thirst. Where a patient lost large amount of fluids through urine, there is that desire for water which is never enough.

2. Excessive Urination: Where a person urinates frequently, it is a sign of diabetes due to increase of sugar in the urine and should see a doctor immediately.

3. Excessive hunger due to diabetes: This is an abnormal hunger which is a reaction in the body due to lack of glucose and therefore starving the body cells. Here the patient is obsessed with food which turn to increase the glucose level, and increases in the body weight.

4. Weakness and Tiredness: The body gets weak and tired, when it feeds on the proteins that is in the body. And not necessary the fats that is been used.

5. Itching: Where there is persistent itching especially around the genital area of the body. Itching can also be experience in the body generally. But where a woman constantly experiences itching in her vagina, she should see her doctor or go for medical checkup.

6. Loss of weight: Where the cell in the body is starved of glucose, the body make use of stored fats to nourish its cell. And when it is done for a long time the patient will begin to experience weight loss.

Here are five top ways for preventing (and even reversing) diabetes.

Five steps in the prevention of Diabetes

1. Consumption of refined sugars should be avoided completely. These are food substance like cakes, cookies and drinking of soda. Also floury foods like pizza, white bread, biscuits, pancakes and many more should be avoided completely.

2. Eating of plenty of organic vegetables and foods like spinach, lettuce, smoothies, brown rice, and high fiber foods like green plantain, to eat with vegetable stew or cabbage stew. Bitter leaf and Ugwu leaf are also good for the cure of diabetes. This can be done by blending, and drinking of the juice.

3. Exercise for about 30-45 minutes every day can help cut down or eliminate diabetes. Back yard garden can also be part of a good exercise for the body.

4. Completely abstain from process or junk food. However, this can be difficult with people of low income as junk food is cheaper out there than the organic foods. But the truth is that we can plant some of these vegetables behind our houses and learn to cook our own food.

5. Vitamins D is another good source of curing diabetes. There are some supplements vitamin pills which you can contact your Physician for the best possible to be taken one a day to help reduced the diabetes.

To the extent that you can do these five things, you can reverse diabetes yourself! Diabetes is not a difficult disease to prevent or reverse because it's not really an affliction that "strikes" you randomly. It is merely the biological effect of following certain lifestyle (bad foods, no exercise) that can be reversed in virtually anyone, sometimes in just a few days.

The truth is we change our lifestyle and diabetes is easily curable if only we can change our eating pattern.

But people are curing diabetes every day. It's simple and straightforward, and when you cure diabetes, you greatly reduce your risk of heart disease, obesity and cancer at the same time. The thing is, no one will cure your diabetes for you. Sure, the drug companies want to "treat" you with diabetic drugs, but you have to keep taking those for a lifetime. They don't cure anything. The only real cure can come from YOU, by changing what you eat and increasing your exercise.

So I urge you to get started today. Just couple of days from now, you could be off insulin, off diabetic drugs and back on track with a healthy life.

What is a Pancreas?

The pancreas is an organ that excretes hormone insulin and pancreatic fluid, which contains enzymes involved in the digestion of fats and proteins in the small intestine. The pancreas, which is shaped like a human tongue, lies below and behind the stomach and in between the two kidneys. It weighs about 100gms, and is made of small units called lobules. Each lobule consists of two groups of cells, the exocrine and the endocrine. The endocrine group of cells is called an "Islet of Langerhans". The beta cells of the islets produce insulin while the alpha cells of the islets produce glycogen, a hormone which does the reverse of what insulin does, that is, causes a breakdown of glycogen into glucose, thus raising the sugar level and preventing it from falling too low.

Health Benefits of Bitter-leaf

Bitter leaf also known as onugbu in Ibos, Shiwaka in Hausa, Ewuro in Yoruba and Oriwo in Ishan. It is good for increasing of blood pressure where it is low, and to prevent diabetes in case of high sugar. It is also good for kidney and liver problems as it serves as a repairer. Bitter leaf contains Vitamin A, C, E, and Vitamin B1 and B2. Bitter leaf when blend with water, you sieve, and drink can have a lot of good effects in the body. So please go ahead and try drinking bitter leaf juice those with the above ailments. It is grown mainly in the tropical area of Africa, but it can be found everywhere.

Unripe plantain as a cure for diabetes.

Plantain look like banana, but larger than banana. It contains Fiber, which help in the prevention of heart diseases, diabetes and certain cancers like colon cancer. It contains Vitamin A essential for healthy eye and normal growth. It also contain calcium necessary for healthy bones, teeth, nails and muscles which help to strengthen the bones. It can only be eaten after being boiled, baked or roasted and can be eaten with roasted peanut, baked fish, stews and vegetables sauce.

The Nutritional Value for an unripe Plantain is an important food to be included in the family diet plan. Plantain is one of the most important healthiest food as it contains a lot of iron, vitamins, less carbohydrates, minerals and good source of food energy. Unripe plantain can be cooked in many ways and forms. It can be cooked in the form of porridge, baked, roasted, boiled, fried and in the form of yam flour. It can be eating with vegetable soup, melon soup, fish, chicken or beef stew. The roasted plantain (bole) can be eating with peanuts or palm oil added with some salt and pepper, it is very delicious. Plantain contains less fat and sugar and should be an added item in the weight loss program since it is good in the lowering of the blood level and the maintenance of a healthy heart, mind and body.

Juice can be made from an unripe plantain, with an added ginger and honey or just drink it on its own as to freshen the body, maintain your shape and makes one look younger. Unripe plantain can also be used as a facial mask, by taken off the peels scrape some of the skin form the main plantain and rub it on your face. Leave it on the face for about 10 minutes and rinse as it will help remove the oil from the face. Plantain can be found in the local grocery shop near you.

Other kinds of vegetables and fruits that are essential for diabetes control are spinach, bitter leaf, ugwu leaf, lettuce, mushrooms, sweet potatoes leaf and the fruit, carrots, broccoli, peas, radishes, onions, parsley cabbage, cucumbers, garlic, beans, cauliflower, asparagus, red peppers and squashes. For the fruits eat lemons, limes, mangoes, oranges, banana, kiwi, almond milk, plum, pear, avocado and all the berries.

For the meat eat plenty of sea foods like catfish, herring, shrimp, oysters, tilapia, salmon, turkey, tuna, chicken, hen, crabmeat scallops and sardines.

For grains eat brown rice, wheat and coconut flour.

For nuts eat peanuts, almond, pecans, flaxseeds, sunflower seeds, cashews and walnuts. They should be unsalted and if possible raw.

CHAPTER FOUR

EXERCISE AS A WAY OF WEIGHT LOSS

Losing weight is a serious business in our today's world with regards to increase in obesity. People are becoming aware of what overweight can do to their body, health and their lifestyles. Losing of weight is good for several reasons. It is an advantage to people suffering from diabetes, shortness of breath, high blood pressure, joint problems and increased in cholesterol. Losing of weight can be done through exercise and healthy food choices which comprised good quality protein food which will help to lose weight quickly, and live a healthy life. Losing of weight can be done by living and eating right.

Basics of losing weight: Consume less calories than usual. Losing weight can be very important, if one maintain a proper diet control and plan. Some people may want to go in for bypass surgery in order to slim down. If not control or eat right, there is the tendency to gain back the weight. One of the best way of retaining weight lost is to keep a healthy lifestyle, eating the right food and doing exercises at least 30 minutes a day. Gradual and gentle stable weight loss for a long term achievement is very necessary. Before going into weight loss exercise, it is essential to equipped oneself mentally towards the journey of the weight loss and lifestyle changes that one is going to embark on. And make sure you are constantly receiving advised from your physician.

Developing a good eating habits, will be a great improvement in one's life, personal contentment gratification, and self-assurance. This brings out the beauty in a person life as a result of weight lost.

For weight lost to last longer, one will have to totally and enduringly changed their eating habits. Where he or she has any medical disorders or ailments, one may want to consult with their family physician before thinking of engaging in a continuous weight loss or any exercise regiments.

Drinking water is one of the best, greatest, and quickest way of losing weight that dieticians recommend. This is essential to burning 100 plus additional calories a day. Every twenty soft drinks one avoids, could average about a pound of losing weight.

Diet

Dietitians and nutritionist are those who work with clients and patients when it has to do with his /her dietary needs. While dieting will decrease the intake of calories, exercising will help burn additional calories. Losing weight through dieting is important if he/she is overweight or obese. When he/she eats a lot more of vegetables, losing weight will not be a problematic factor as it contains less calories and carbohydrates.

A healthy food choice decreased in calories diet with reasonable fat is suggested or recommended. Different kinds of fruits are recommended as part of weight loss diets as a healthy means of controlling starvation and to provide the body with the necessary nutrients and vitamins that the body needs to function daily. But the fruits should not be eating too much because of its sugar contents.

Exercise while dieting: Losing of weight is about reducing the amount of calories consumption while one adds to the amount of calories one burns. One must first see his or her physician on the amount of weight to cut before starting the fitness exercise. A diet that may work for one person may not ordinarily work for another person. It is essential to eat a healthy breakfast in the morning as it is one of the most important fundamentals of a healthy diet and a major attribute to weight lost. Diets when it is adhered to very strictly or meticulously will end up in weight lost as a result of caloric curb or control. Dieters who fail to implement or embrace healthier exercise and eating habits will regain the weight and possibly extra.

Several of the dieters, who do not exercise on a regular basis regain the weight in less than one or two years. And this help the body to store fat and possibly the risk of acquiring heart disease. Eating a stable, modest-ration of food three times a day with the main meal at noon further operate in preventing overweight. Crash diets should be prevented, as it only helps in adding weight to the body.

Decreased of belly fats and having a healthy stomach area can be attained by constantly exercise practices and food choices. Performing physical exercises everyday will assist in sustaining a fabulous waist line and firm muscles. Doing exercises will bring in additional metabolic in the body to porch out extra calories of what the body takes in each and every day. The stomach area around the waist line can be improved by doing stomach fat lessening exercises.

Procedure/technique

One ought to lie down on the back straight on the carpet for the exercise and place both hands on each side of the body with the palms facing down. While in this position, take a deep breath and get ready for a walkout. Slightly lift up your legs so that both legs are upright to the floor and foot is facing upward to the ceiling. Raise your hips completely off the floor and shove your hips up for about 2 or 3 inches. Grasp the position for a few seconds and again elevate the hips to the earlier position. Do the exercise 10 times each for a couple a day. After doing the exercise as required, place your feet on the carpet and take a deep breath by sitting up and in a comfortable position. Stretching can be done to activate the body. Make sure you lie down with your back always flat on the carpet.

Breathing

In doing breathing exercise, exhaled air gradually by raising your legs and hips above the ground and inhaled air gently by going back to the earlier position. It eases the rich supply of clean oxygen to the blood vessels. The air should be inhale when the stomach is in the state of a long drawn out, and then exhale air when the stomach muscles cuddled to burn the stomach fats at the time of workout. Inhale and exhaled also expedites the removal of contaminations that are present in the body through sweat.

Defenses and Effects

While it is true that the breathing exercise can be done easily in the house, it is advisable to do it outside for fresh air and clean supply of remarkable oxygen to the blood vessels. The use of jogging pants and shoes are recommended during the exercise as they are weightless and stretches. They prevent the flow of blood as it helps to permit or allowed physical movement. Decrease in stomach can be of advantage only if the exercise is done on a frequent basis for about two to three months. Some nutritious rich drinks and water should be taken after each exercise to freshen up the body. Avoid eating solid foods before or immediately after the exercise.

Sit Ups

Sit ups are good exercises that aimed at decreasing the stomach fat and can be done in minimal effort and time. With will self-determination, will power, and encouragement, it is easy to lose fat around the stomach area.

The Procedure

Lie down flat on your back on the carpet, raise your legs up straight and knees at 90 degrees and feet on the ground. Hands kept near the ears or below the chin. With eyes closed, take a deep breath as to stimulate blood circulation and to prepare the mind for a physical walk. Try to move up the upper portion of the body while the stomach is still on the floor, and then do a twisting motion. As you do the sit up, emphasis should be on the stomach to strain the stomach muscles to spend energy. Do this continuously with a break in between each day for a better result.

Precautions

This exercise can be done both indoors and outdoors for fresh air and clean supply of natural and odorless oxygen to the blood. Jogging pants or shorts and good shoes should be worn to avoid blood clot and to permit physical movement. Breath air in and out during the sit ups exercises.

Effects

It Improves suitability of the legs and raises the rate of metabolism in the body with increase in temperature. This exercise is done to improve the overall appropriateness of the body as the blood circulation is given a new life now and again with fresh clean air.

Stomach Twisting

This exercise is good for the elimination of all the numerous fats stored in the stomach by eaten junk foods like hamburger, pizza, fried foods, French fries, and to provide a firm opposition to action of digestive enzymes and acids to the body. These acids and enzymes are directed to a cover of the skin tissue all over the body especially around the stomach. And so by doing the stomach twisting exercise, it can help cut off excess fats in the stomach. To have a drastic or quick results, it must be done on a continuous bases to eliminate all the problems of stomach fats.

Stretching

Stretching should be part of a daily routine. Just a few minutes of concentrated stretching can help you feel better and loosen you up; and prevent muscle injury too! Popular walkout like yoga, Pilates, and tai chi, can improve muscle strength and overall cardiovascular health. When stretching is done in the right way, there is increase of blood flow through the body. This, increases strength in the muscles, build tissue elasticity and improve the range of motions in your joints. Whether you mainly sit or stand at your workstation, it is recommended that you find time to stretch and move about, or take small breaks to give your muscles a chance to recover and release some tension too.

Stretching Tip:

Poor blood flow and muscle tension may lead to headaches or discomfort. It can lead to stress build up, but one way to relieve stress is stretching. Stretching the right way increases blood flow through the body and physically releases stored tension in the muscles to help you feel better.

Keep it fun by joining a group

If you are really struggling to stay motivated, your walkouts are probably not much fun. Consider changing things by signing up for a class like dance or martial arts where you can learn something new. You can even recruit a friend to compete with you for a change. It helps from keeping the same routine every day.

Make yourself Number one

Many people tend to make themselves the last priority, when in fact they should put themselves as number one. Walking out and getting in shape will ultimately give you more energy that will enable you to take better care of your responsibilities. Your family and friends will appreciate you taking care of yourself first!

Exercise for the whole family

Health and fitness are important for persons of all ages. Arranging up a lifetime of fitness and exercise can be one of the paramount gifts you can give to your child, and also as an advantage to yourself of being around longer to see the children grow to a middle age and even old age. Are you ready to take your first step to a more vigorous lifestyle, by partaking in physical actions and an adoration of exercise and keeping fitness is a mighty way of attaining an ideal result for the whole family.

1. Family Hikes

Family hikes is lovely when done outdoors and often help to advance good exercise habits in kids as well as an attraction for nature. It is best to figure and develop a lasting habit and love for physical actions instead of a dislike by forcing children and family members to exercise. Hiking offers a great prospect to explore tracks, and getting the whole family members take turns in choosing what is the best paths for them. It is a funny practice by inspiring kids to find out interesting plants and wildlife, counting flowers, singing songs and enjoying a picnic along the way, as they hike and they will always be looking for more, anytime the occasion arises. Parents should consider taken the whole family to both a fabulously charming national and state parks that have hiking tracks as well as campgrounds, as a great family trip to the outdoors events.

2. Indoor Climbing Gym

One thing that kids love to do is to climb on chairs or on tables. And having an indoor climbing gym that has a rock climbing wall that is safe, is another good way of having fun and at the same time getting fit. There should be binds and lifters to assist those who will slip, to avoid the children from injuries. Adults too are not left out. They too can benefit from the climbing as it is another way of keeping fit.

3. Competition and Charity Events

There are many ways of keeping fit. These could be by joining a group and competing with one another or involved in the fitness exercises or walking club. This is great as several people, young and old, are inspired when they compete as a group. Charity events can be very good as it offers a great way to give back, and through walks, bike rides or a run, it gives an additional meaning to fitness. As a family making an arrangement for fitness exercises will be the best thing. This will raise everyone spirits in going all out in losing weight.

4. Disc Golf:

Disc golf is another new sport that everyone in the family will appreciate. It is similar to traditional golf and it is easy to learn with starter discs. Disc golf can be played by individuals with varying skills, using the upper and lower body strength and good for blood circulation. It decreases the risk of serious injury, making it safe, fun and family-friendly.

5. Yard Games

Yard games provide a great way for fun and physical activities into one's daily life. Meditating, kickball, flag football, whiffle ball, basketball, or picking a ball and throwing it at one another is a great way for the family to enjoy the exercise, getting fit and spending time together, attaining a fitness goals, and promoting a long lasting love of a healthy physical exercises.

The secret of looking younger

Being healthier and having more energy requires a small waist which improves your looks and cuts your risk of diabetes. Lose your belly and strengthen your core, will improve in acquiring a complete and healthy life. Excess weight anywhere in the body is bad for one's health, and belly fat is harmful as it tends to settle around the essential organs like the liver and it blocks the body's ability to use insulin, the hormone that regulates blood sugar levels. A modest exercise about 30-45 minutes a day can improved insulin sensitivity, cut down on liver fat, and reduced swelling in the belly even without changes in one's diet. Exercise also reduces the *size* of abdominal fat cells that are under the skin, being another risk factor for diabetes. So get moving today! It will improve your health, not to mention your love and sexual life.

The History of Belly Fat

Since prehistoric times, men and women have sized up possible mates based on their waist-to-hip ratio. A bulging abdomen on a woman signifies that she was pregnant and not marketable. On the other hand, it could be that the woman might be infertile because of a hormonal imbalance. Belly fat in a woman was a sign that she would not be an ideal companion. While men, a large abdomen suggested low testosterone levels. Excess belly fat usually converts testosterone to estrogen, which can lower a man's sex drive. One is programmed to respond to his or her companions based on their body's reproductive prospective. Reducing of belly fat is good for one's health and self-assurance and a more confidence.

Measure Your Belly Fat

To find out if your belly fat poses health problems, first calculate your waist border.

Grab a tape measure and follow these steps:

Put the tape measure around your waist at your belly button and write down the measurement. Check out the ideal measurement for one, s height and divide the height in inches by two. For example, a height of 5'4" (64 inches), then your ideal waist circumference should be 32 inches. Women's waists line should measure 32.5 inches; while that of men should be 37 inches. Slight increase in the waist line can make a large change. Reducing waist border is "the single most obvious thing one can do for safety and to avoid any sickness.

Lose weight by trying these 5 flat stomach exercises to trim and tone your abdominal muscles:

Scissors Kicks

This fast-stride move will strengthen your lower abdominal and inner thighs.

Step 1: Lie flat on your back with your legs stretched. Place your hands flat on the ground with palms down, holding them under your butt slightly to create power.

Step 1

Step 2: With your feet raise up about a foot off the ground. Kick your feet back and forth from side-to-side, emulating the movement of scissors. Focus on your lower abdomen as you complete 12-15 reps. (making sure you breathe steadily throughout the walkout.) and then rest for 15-30 seconds and then continue.

Step 2

Scissors Kicks

Weighted Oblique Twist

The weighted oblique twist will strengthen your oblique's, which in turn will help strengthen your core and lower back, improve your posture and help prevent injury from any bending or twisting movement.

Step 1: Sit on the ball, holding a dumbbell just under your chin. Walk your feet forward until your hips are slightly lower than your chest. Keep your abdominal tight and chin up. Turn to one side to begin the exercise.

Step 2: Slowly twist your torso to the other side through a count of 10.

Step 3: Hold and squeeze for 2 seconds at the maximum tension point.

Step 4: Then twist to the other side through a count of 10 seconds. Repeat three times without resting.

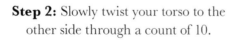

Side Plank

This move will strengthen your abdominals, back and shoulders. Be sure to breathe normally as you perform it.

It will test your balance skills, but most importantly, it will stabilize your torso, helping you do simple tasks, such as lifting items and opening doors. Doing side planks will also improve your performance in any sport.

Step 1: Lie on your side with your legs stretch and left arm resting in front of you. Support your upper body with your right forearm against the floor.

Step 2: Breath out as you lift your hips off the floor, balancing your body weight on your right forearm and outer edge of your right foot.

Step 3: Hold for 30 seconds.

Step 4: Switch sides and repeat for a total of 4 times.

Flat levels

This move will tone and trim your abdomens as well as strengthen your back and pelvis.

Step 1: Lie on your back flat, stretched your legs and hang your feet just above the floor. Rest your arms at your sides with palms against the floor.

Step 2: Exhale and tighten your abdomen as you lift both legs, creating a 45-degree angle between your legs and the floor.

Do not allow your lower back to roguish away from the floor.

Step 3: Continue to breathe normally as you hold the move for 5 seconds.

Step 4: Lower your legs to the floor.

Step 5: Rest for 20 seconds and repeat three times, for a total of 4 repetitions.

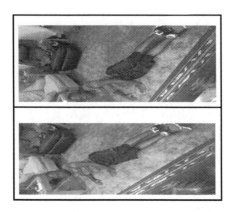

By my nephew Princely Besong

Oblique V-Up

The Oblique V-Up will strengthen your lower abdominal muscles. Try it out!

Step 1: Lie on your side with your legs stretched and left arm resting in front of you. Prop up your upper body with your right forearm against the floor.

Step 2: Raise your legs, keeping them together, about four to six inches off the ground. But do not raise your upper body to meet the movement. Focus on your oblique's as you raise the legs.

CHAPTER FIVE

BEST EXERCISE TO REDUCE BELLY FAT BY SALLY IYOBEBE

Hip Raise Exercises

This is one of the exercises that helps in decreasing belly fat of which it is best if one wants to maintain a flat waist and to build muscles through glutes. When exercising everyday it will help one burn more fat quickly.

Procedure

One should lie down flat on the back, with your knees bend and arm stretch out to about 60 degrees. Squeeze your glute tightly, take a deep breath and raise your hips in such a way that your body aligns with your knees to your shoulders. Do this for about 5 seconds. Repeat the exercise for about 5 sets, then a rest for about 2 minutes between each sets and making sure your back stays flat on the floor.

BEST HEART RATE FOR WEIGHT LOSS

The actual zoning, tends to be at an intensity of 60 – 80% of your maximum heart rate. Where one has been lazy for a long time, then exercise at an intensity level lower than 75% of your maximum heart rate and then gradually increase the intensity. You should always stop and assess how you are feeling during your walkout before you continue. While walking, you can sweat and breath harder than normal, then you are walking at the right moderation, and you do not have to pant for breath. You also need to sustain it for 30 – 60 minutes. As you become more fit, it becomes easier to exercise in your target heart rate (THR) range. But, you need to occasionally re-calculate your target heart rate, as it sometimes decreases due to constant physical activities.

RATINGS OF PERCEIVED EXERTION (RPE)

Another way to assess the intensity of your walk out is by using a scale of 1-10, called the Rating of Perceived Exertion (RPE) scale. The scale does not require measuring heart rate. This can be easier to use for beginners who are not using the gym. If you are using equipment in the gym or have a heart rate monitor when doing outdoor activity, it is simple enough to use target heart rate as your intensity guide. The best exercise is the one you enjoy most and are able to do safely and effectively, and suits your current fitness level. Try walking, jogging, running, cycling, swimming, Nordic walking, dancing, jumping jacks and other activities (check here for weight lost exercise activities). If you want to get seriously fit try running, have a running plan as a new comer. If you have joint or back problems, try doing some cycling. Ask your physician about the right exercise program to follow, which will help you lose belly fat and improve health.

RUNNING FOR WEIGHT LOSS

For one to have a great body, he/she must avoid high intake of calories, and also do some exercise. Dieting alone will not maintain the kind of body you want, so nothing should stop you from exercising. You can start by running slowly for about 5minutes, stop and walk and then run again, until you get used to running for about 20 to 30 minutes a day. But it is recommended that you seek your doctor's advice. Do not bother yourself with weight lost pills because it will never help you or do any trick.! **NO! THEY WILL NOT!** Do not be deceived. Those pills will never work!!! So **do not** go out to spend your money on a product that will not help you lose weight. Your best bet is to exercise, eat plenty of vegetables, less fat and carbohydrates, drink plenty of water, no sodas and then do some fasting.

While studies show that calorie restriction (dieting) is effective, in terms of weight loss, they also show that they are almost useless in isolation in maintaining this weight loss. Whatever you think now, a few days into your diet you will discover that self-denial and abstinence is not good for the body over a long period of time. Exercise in combination with dieting is the best way of keeping the weight off the body.

Dieting alone will help you lose weight, together with a substantial sum of muscle. However, you **DO NOT** get a toned, shapely body thrown in. The only way to get a healthy looking, slim and toned body, is to add exercise to your weight loss plan.

Try not to desire for the same kind of foods you were eating before you started the regular exercise. You must discipline yourself so that you do not fall back.

Stronger heart and lungs. As your heart and lungs become stronger and more efficient, you are able to take in more oxygen with each breathing exercise., improved lung function and help improve lung conditions like asthma. Seek your doctor's advice first before you embark on any exercise.

Running can help protect your joints, and delay the onset of arthritis. It is a great technique for strengthening bones in one's legs and hips. Proper shoes and clothing are necessary to prevent injuries.

Improved skin. Running can improve one's mood, and makes one's feel relaxed, content, calmer and less anxious. Running is very satisfying as it makes one have a great sense of accomplishment.

WHY RUN?

Running is the cheapest means of exercising. It does not warrant one to be a registered member in any gym club. All you have to do is just go to the park and run or within your neighborhood or in the school stadium. It is more refreshing being outdoor and getting all the fresh air when you are exercising. Proper dressing is needed for the exercises. You can exercise on a treadmill in your house, or go cycling if you have a bike.

If you cannot run due to overweight, you can find a good exercise that is good for blood and fat circulation to cut down on those extra pounds in your body.

RUN, JOG OR WALK

What's the difference? Walking is when one foot is always in contact with the ground. The second foot touches the ground before the first foot leaves the ground. In Jogging, you are either a running or walking. You may be a slow runner or jogger, but you are still running. Walking at a pace towards the upper end of the natural range of walking is known as power or speed walking.

You either go for running at a steady step, or you can do all by walking, jogging and running at intermittently. When running, jogging or walking, start slowly for some couple of days, before going into the fast way. You WILL still burn calories going slowly, if you exercise long enough. If you exercise at a moderate to high intensity, you are on the right track. You can play some good music while running as most people's ability to run well relies on music. I suggest you get an iPod or other mp3 device, and run to the rhythm. Once you lose belly fat, you want a body that is strong, muscular and well defined.

Anti-Stress Remedies That Work

What we usually refer to as stress is a physical response to all of life's demands.

Busy schedules, challenging relationships and anxiety about health or finances can all contribute to the physical and psychological symptoms of stress, such as body aches, and lowered immunity. It is important to manage stress well, through the following methods:

Good Nutrition

Certain vitamins and minerals can be called into action when the body is under stress. For example, vitamin B6 works good for the body and is found in seeds, leafy greens, beans, egg yolks and fish. Stress-busting minerals like potassium and magnesium(chocolate), are muscle relaxers. To avoid the sugar high that accompanies chocolate, try nuts, spinach or beans, which are also rich in magnesium. Potassium, bananas or whole grains are also good choices.

Exercise

Exercise is one of the most effective stress remedies because it raises endorphins and keeps your energy levels up. Exercise keep your muscles from becoming tense, which is a great way to blow off steam when things in your life seem out of control. Taking a brisk walk, stretching and even light weight lifting are all effective ways to manage stress with exercise.

Yoga and Meditation

Physical yoga postures are great for stress management, because yoga can be a powerful tool for calming the mind. One of the goals of yoga is to prepare the body for meditation, and the purpose of meditation is to create stillness in the mind that allows you to live in the moment. In this state of relaxation, you will have less anxiety and fewer symptoms of stress.

Aromatherapy

The area in the brain that processes aromas is also responsible for emotions. Scents can have a direct effect on how you feel. Calming aromas that can help with stress management are purple, geranium, cedar and jasmine. Sandalwood, peppermint and lemongrass have a calming effect as well. The most benefit from aromatherapy, is to use pure essential oils. Aromatherapy oils can be diffused, massaged into sore muscles or added to unscented body products or bath water.

Herbal Remedies

Herbal remedies can be an effective way to deal with the anxiety that often accompanies stress. Herbs that are used for this purpose include ginseng, chamomile, valerian, passionflower and lemon balm. You can buy these as capsules, teas or essential oils in most health food stores as well as traditional pharmacies. All of these methods are simple and effective tools for managing the stress of everyday life. If you are under severe stress, however, do not be afraid to talk to a counselor or medical doctor for additional help.

CHAPTER SIX

ARTHRITIS

Arthritis has no cure, but those suffering from it, have a powerful, and cheap way of managing their aching joints. There are 5 best exercises for people seeking joint pain relief, and are easy tips to get you started. Make your walkouts be part of a fun and stress-relieving, to ease your symptoms and help you lose weight quickly.

To prevent arthritis or any incident that can cause the risk of one having arthritis, is to maintain a healthy weight necessary in the prevention of arthritis. Stop smoking and eat a healthy diet that is low in sugar, alcohol and purines. Avoid eating fast food, greasy or oily food, French fries and pizza. Wear comfortable shoes as high heel shoes can cause arthritis. Physical activity like swimming and a walk out are the best medicine for arthritis pain relief. Exercise, such as walking is just as effective in reducing knee pain. Exercise can decrease pain, particularly for people with osteoarthritis, the most common type of arthritis. Regular walkouts may prevent sore joints and stop arthritis from getting worse in the long run. physically active people seems to have a higher quality of life and are less likely to become disabled or have days with lots of pain.

Walkouts keep off the pounds. Obesity brings double trouble: It increases the risk of arthritis and makes its symptoms worse. How much you exercise is up to you, your schedule and what your body can handle. For arthritis joint pain relief, just 20 minutes three to four times a week is enough to make a difference.

There is no time to waist. You can break your exercise into two 10-minute intervals in the day.

If you are afraid your sore joints are too swollen to exercise, schedule a doctor's visit.

Sally Iyobebe

Before you start any exercises, you will need your physician's approval.

Your walkout should be challenging, but not painful enough to cause injury. You will be over doing it if you have sore joints or muscle pain that continues for several hours after exercising or if pain is worse the next day. You should shorten your walkout or do it more gently. Get your vitamin D supplements and stay hydrated by drinking a lot of water.

Some exercises for those with arthritis.

To get started, see these 5 arthritis exercises that can help ease your arthritis symptoms:

1. Walking

Walking is very essential and help to burn down fat. It is the quickest way in reducing and burning calories in the body. As you walk daily for about 30 minutes, it will help in weight lost and boast one immune system and moods. It strengthens one muscles and take away pressure from one's joints and reduces pains. It is always necessary to walk for about 3 to 4 miles per hour each day in a week. To make it simpler and enjoyable, use a CD player when walking or just sing. It is also advisable to walk in a group as it will help one another accomplished weight loss exercise.

2. Water Exercise

It is one of the best exercises in terms of weight loss. It can burn fat quickly and ease arthritis. High calories can be burn within a limited time while swimming. It helps relaxes one's muscles and reduces pains. It gives one the kind of balance that is needed for the body. When in the pool, stay on the lower side where your feet's can reach or touch the floor. Drink a lot of water before and after you exercise in the pool. One can do some physical exercises while in the pool like jogging, playing of basketball or volley ball, and hide and seek exercises. All these help in cutting down fats from the body.

3.) Yoga

Yoga exercises helps in burning calories and makes you a super slimmer person, changes your body and mind. It is a stress buster, helps you recover from a long walk and it eases aches and pains. It improves flexibility and makes you even stronger. If you want to melt away some pounds of fats from the body or abdomen, doing some yoga exercises will help.

4.) Tai Chi

Before you start doing this exercise, starts with some push up, breathing, balling or standing exercise. By doing tai chi exercise, the best is to circle your hands around your upper body. This should be done slowly and in a vertical position, up and out from the body while keeping the same distance apart, before bringing the hands down and inward. This exercise is good because the whole body is involved in the process.

5.) Jogging

Jogging is one of the most popular and oldest forms of aerobic. It is always good to start slowly, with a brisk walk in between. A good shoes and sports clothes are necessary when jogging. Jogging can be done either indoors on a treadmill, outside in a park, school stadium, or around the neighborhood. It helps in burning fat quickly. Jogging prevents diabetes, infectious diseases, cancers, heart diseases, and high blood pressure. Jogging, is good mentally as it reduces stress, emotions, and body fat in weight lost. Now is the time to go out and jog as it is the perfect answer for weight lost and arthritis.

CHAPTER SEVEN

FASTING TO LOSE WEIGHT

What is Fasting?

Fasting is abstaining from food, drink, or sex to focus on God for prayers and spiritual growth. This is a period we humbly deny ourselves something of the flesh to glorify God. Fasting enhances our spirit and goes deeper in our prayer life. Fasting could also be a time when one goes without food to pursue and focus on something in the inner man. In 1st Cor. 9:27, fasting helps subject our bodies to our spirits. In prov. 25:28, fasting is disciplining the body, mind and spirit. In Gal. 5:17 Fasting is subordinating our flesh-desires to our spirit-desires. In Mt. 6: 33 fasting helps set the priorities in our lives. And in Ps. 63:1-2, Fasting is longing after God.

Too often, the focus of fasting is on the lack of food. However, the purpose of fasting is to take our eyes off the things of this world and instead focus on God. Fasting is a way to demonstrate to God and to ourselves that we are serious about our relationship with Him. Though fasting in Scripture is almost always a fasting from food, there are other ways to fast. Anything you can temporarily give up in order to better focus on God can be considered a fast (1st Cor. 7:1-5). Fasting should be limited to a set time, especially when the fasting is from food. Depending on how you want to be closer to God and for him to answer your request, you can fast for an extended periods of time without eating. After all Jesus fasted 40 days and night without food and water. And today a lot of prophets are doing a lot of extended dried fasting to seek the face of God and it is actually working in helping to promote the gospel and their ministries.

Start with a very short-lived fasting program before embarking on a longer program for weight loss. For first timers, start out fasting for only a portion of the day. From 12 midnight to 12 noon the next day. Gradually you can then progress into a longer period of fasting. It is necessary not to subject your body to a longer period of fasting without adequate preparation time for adjustment.

Prepare your body to undergo a fasting program for weight loss by limiting unnecessary food habits, such as the intake of alcohol, sugar and caffeine at least 1 week prior to starting your fast. You should refrain from and limit the use of nicotine and avoid eating red meat if you can. One of the primary purposes of fasting is to rid the body of toxins, so it will be helpful to get yourself in the habit of avoiding these substances before you even begin.

Allow yourself to eat only one meal per day while fasting. The best time of the day to have this meal is usually about 3 pm in the afternoon, as this is the midpoint of the working day for most people. The meal you have should only be comprised of water, tea or juice and some form of raw fruits or vegetables to ensure maximum detoxification and weight loss.

Do minimal exercise and avoid strenuous activity during the period of fasting. This is because your body will not be physically capable of performing any form of strenuous exercise due to the physical effects of fasting. Denying your body food and water temporarily decreases the amount of stamina your body is able to expend.

Re-check the amount of food you eat after you finish your fast. It is naturally to want to eat heavily after completing a fast, but it is equally important to slowly increase your intake of food over a period of days to allow your digestive cycle to return to normal. If you weigh more than 50 pounds' overweight, I will strongly encourage you to consider fasting as a means of cutting down dangerous pounds very fast off the body. I would implore you to set aside everything else going on in your life and take immediate stock of whether fasting is for you or not. And it doesn't have to be a very long fast either.

Another good fasting is water fasting that can help you cut down quickly in two weeks and you will loss as much as 20 pounds. Fasting for as little as three to five days can take as many as ten to twelve pounds off your body, thus reducing the risk of chronic disease. Try to stay free of junk and maintain permanent eating-habit lifestyle changes. There is nothing more powerful than fasting to lose weight fast and getting rid of excesses, improve physical, mental, emotional and spiritual health. All you have to do is to discipline yourself and make a plan to stay fit by exercising every day and eating healthy food.

Eating too much food is just like an addiction to smoking and drinking. And so you will need a strong determination to stay free from this addiction. Due to poor eating habit over a long period of time, your body can become addicted to all kinds of bad food and this need to be stopped and cleansed. And to cleanse your body from this poor food habit, you need to cleanse your system with plenty of water and juice, as fasting can purge and cleansed the body and weight loss quickly. This help to take off so much amount of fats from your body in a very short time. You need to motivate and discipline yourself. You need to love yourself first before loving someone else. Just think of how you will feel if you are able to flush out those excess fat from your body. Do not give yourself anytime to relax until you make sure you have taken all those excess fat from your body. Learn to treat your body in a special way. For ladies, treat yourself like a princess or queen, and for the men, like princes and kings.

The Purpose of Fasting.

In Luke 4:1-14, Jesus fasted for forty days without food. It was during the end of the fasting that the devil came to tempt Him and commanded Him to turn the stone to bread, if He Jesus was actually the Son of God. The purpose of Jesus fasting to mankind is for us to become more active and productive. It helps us focus on spiritual problems. Therefore, fasting is more than abstaining from food and wine. Prayers should have accompanied fasting as it is an essential part of the fasting, couple with sincerity and honesty. There are several benefits that comes with fasting as it helps and guide during times of sickness and for blessings. It helps us to overcome weaknesses, comfort us in times of travail, and bring us closer to Him as we get to know Him better.

When fasting, have a cheerful face and countenance. Encourage members of your immediate family and friends to fast. Fasting can help you save a lot of money which can be used to help those in needy and to advance the gospel. It gives you are alter most peace, health, emotions, take out bad habits and spiritual life. It also helps in improving our life's, character, control our cravings and lust for the things of the world. We have self-control, overcome temptation, and develop faith in God and ourselves. During fasting, God reveals Himself to us, as according to Isaiah 58: 8-9, "when you fast and pray and call on God, he will answer as He is always there for you". Regular fasting cannot kill the body. We may lose weight and strength during prolonged fasts but the spiritual breakthroughs in moving closer to God are too precious compared to anything we can lose.

Dramatic and Rapid Weight Loss Fasting

The best way to lose weight rapidly is by fasting with water and juice for seven days. I have done it and I encourage you to do it. You will not die as a lot of people will say, I cannot fast even for one day. But if you give yourself a trial, you will succeed and become used to it. And believe it or not you will lose about ten pounds. But I will advise you take it gradually by starting with one day fruit or water fasting and you will lose about two to three pounds which is very encouraging. From then on keep on doing it until you get your desire results. Because of too much fat in some people's system, they might need a lot of water to help flush out the excess fat from their system. And the end result is beautiful and for me I do not want to stop but continue to stay fit and healthy.

During those first seven days of quick weight lost fasting, your body will have scrapped all swelling that was accumulated in your skin, especially if your diet included substantial amounts of salt. On days three and four, the massive fat burning continues as the body get rids of all abnormal fats.

After seven days, water fasting produces an ongoing healthy weight loss of approximately one-to-two-pounds per day. In other words, after your body completes the initial heavy elimination of toxins, usually in nine-to-eleven days, the weight loss will slow to one-to-two pounds daily. One pound per day is by far the most common, based on my experience.

In 2012, I did a 40 days fruit fasting and I lost a massive 70 pounds weight in 6 months. The result was about 'one-pound-per-day'. When I travelled to Nigeria for my daughter's wedding in August 2013, they could barely recognize me because of my weight loss. The only way for you to know how much weight your body will lose through fasting, is to go ahead and try it for yourself.

You might want to try some juice fasting and see the result for yourself. you will discover that you will lose about three-to-five pounds per week on an average. But after that I encourage you to drink a lot of water and exercise to flush out the juice content in your system. This is because juice has a lot of sugar which is not good for those with diabetes. I did it with light exercises because I was determined to lose that much weight 70 bounds and I am happy I did. I am ready to work with anyone who is ready to lose weight but they must be ready to discipline themselves without which it cannot work.

Believe it or not, you cannot keep living your life like somebody in the street. Ladies, you do not want to look as if you are your husband's mother each time you go out with him for an occasion. Surely, you do not want that. And do not deceive yourself that my husband likes me the way I am. You know for sure that is not true either. Try to take care of yourself by minimizing the kind of food you eat. Also try as much as possible to do some exercises every day for about 30 minutes to maintain your body after losing weight. And depending on your body type, weight lost may be quicker or slower. But do not give up for it will help heal most of the arthritis, and other bodily pains. The question is how can one or what other way can one attain a quick weight loss within the shortest time and with a large number of pounds lost, while it is true that there are numerous ways of losing weight, the quickest way to take off all the dangerous pounds is by fasting.

Light Exercise

Another good means to capitalize on quick weight lost while fasting is to include some light exercise daily for at least half-an-hour. Light exercise can be done by going out for a light walking, swimming or slow, and long-distance bike riding. Another great way of fasting is by stretching because it releases toxins trapped throughout the body and thus facilitates weight loss and expedites the cleansing process. If you are into weight lost program, keep the weight lower than you usually would to reduce force. The exercise should be comfortable and easy. Do not have a hard walkout while you are fasting. For it is dangerous and can cause fainting or dizziness of the body. A nice 30-minute stroll, swim or light weight training routine will be more than enough. Always be in constant touch with your physician in case of any negative outcome.

Do not allow your mindset in making you believe that it is never possible to lose weight. Do not give up hope. Just believe and have faith in yourself that you will make it no matter what. Say to yourself that it is high time you are actually going to lose all the fat in your body and to improve in your health. Do not let your mind deceive you into thinking that "nothing will work, that you will never make it; that you are not good enough to lose weight. They all boils down to signs of fear, anxiety, laziness, and making you think you are not good enough to do an exercise. But today I say to you, you can do it. Cast out that spirit of fear and lies away from your mind. The truth is, the fact that you are overweight does not mean that you are weak, or less important. No. It is how you feel about your weight that I am talking about. Being overweight should not be a path in your life. If you know being Obese is a problem in your life/health, and it will affect your sexual life, then do something about it quickly before it gets out of control. I know you have the power to do so and yes you can do it.

You have the power to decide, right here, right now to determine that you are going to give it your all. You are not going to stop until you have lost every last ounce of excess weight off of your body and reclaim the finest health that is rightfully yours. There are nothing like slimy pills, diets or cures. The truth is they just do not exist and if it does it is just for money making of some celebrities. All we do need is the ability to make a firm commitment with ourselves about the quality of lives we wish to lead. All those advert about slimming pills does not work.

It has come to the point where being lean, healthy and free of junk food becomes more pleasurable and desirable than sticking to that cheeseburger, hamburger, cakes or pizza in our mouths. The companies making all those junk foods might not like it but it is the fact. You can always feel good with yourself when you feel thin. Nothing in the world tastes good as when you are light and thin and you can stand on your feet and jump. The only thing that you need for quick weight lost is fasting. For people with more than 30 -40 pounds and above overweight, you are strongly encouraging to consider fasting as a means of getting rid of the dangerous pounds quickly and faster off your body.

We all have tried dieting in many different ways and it has failed us in several occasions. I in particular tried dieting and within weeks I gain back the weight in excess. And that is why I am advising you try fasting without food and see the difference in your life style. It is just like someone who drinks alcohol or someone who is an alcoholic. As long as he/or she keep drinking alcohol, he or she will never be free from alcohol addiction. This goes for smokers of cigarettes, marijuana, cocaine etc.

In most cases, when a person has nothing to eat, fasting is the smartest thing they could do, especially in the third world countries. It is not in America where food is almost around or nearness to us, obtaining it with very little effort required. Processed junk and fast food is readily available, as most people are not ready to cook due to laziness. This is because cooking requires preparation, time, and the washing of the dishes and pots. The truth is nothing good comes easy. When you want good food for good health, you should be able to cook your own food and stay free from junk food. Go for fresh vegetables and meat and your life will never be the same again.

How fasting works

1.) It decreases caloric intake. To lose weight, you need a caloric deficit. No quarrel about that.

2.) It increases fat rust while sparing lean bulk. Since what we are trying to do is lose fat and not only weight lost, the fact that fasting increases hormones that preferentially burn fat and decreases hormones that inhibit fat burning is extremely desirable.

3.) It improves obedience. Fasting can be an extremely tolerable way to diet, especially when compared to outright caloric restriction. If fasting is easier for you than trying to laboriously count calories, fasting is going to be the more effective in weight lost, then go for it.

Fasting is an effective way to lose body fat. It is not the only way, and it is not "required" for original weight lost, but those who believe in good health and life style have found it to be very helpful and by trying it up for yourself it will prove to be the best. If you are looking to burn your fat, buy a copy of this book and call me, and I will be glad to fast with you, if you are really serious in weight lost. But you need to work hand in hand with your doctor for personal observation in case of any side effect.

Medical Fasting

Medical Fasting is an order from the doctor or physician as it is of benefit and necessary to fast before any surgery can be perform. This is done to avoid the body to digest any food while undergoing a slow breathing process of the body either mechanical or chemical before the sedation of anesthesia.

Doctors will sometime require a patient to undergo some fasting for the accuracy of the readings of a medical test in order to gesticulate the blood sugar level or cholesterol level or high blood level. But before any fasting is done for a long period, it must be under the supervision of a physician.

In most cases, Medical fasting can help in weight lost, reduced the pain from arthritis, high blood pressure, stress level or depression. It also help to control insulin in the body. However, those with very serious medical conditions, may not be advisable to fast. But if you can, it will be better, but with good nutritional food before and after the fasting. It helps to reduce the amount of calories in the body because excess calories is bad for the body. That is detoxification of the body by eliminating of toxins through the liver, colon, lungs, kidney, skin and lymph glands.

Also during fasting a lot of healing takes place as energy is moved away from the digestive system due to its lack of use, and gearing to its metabolism and immune system.

To sum it up, medical fasting helps to make you look younger by cutting down excess fat from the body and help you live longer, increased the productions of hormones, add to a good protein that improve the immune systems and anti-aging hormone is produced. Therefore, medical fasting is good for the body since it gets rid of the waste metabolic that stores in the body in the form of fat, and it repairs and heals of all the destructive organs during fasting.

Some doctors will advise on clear liquid, rich in carbohydrates few hours before the commencement of the surgery. This is to prevent pulmonary passing out of waste from the stomach during the anesthesia or when the surgery is going on as a way of reducing the stomach contents volume that may cause damage or death. In some cases, doctors may require the patient not to have any food after midnight as part of the cleansing process of the stomach to avoid vomiting.

Fasting for political reason/hunger strike.

A hunger strike is a method of abstaining from food as a way of political protest, or to provoke feelings of guilt in others, usually with the objective to achieve a specific goal, such as a policy change. Most hunger strikers will take liquids but not solid food.

In cases where an entity (usually the state) has or is able to obtain custody of the hunger striker like those behind bars, hunger strike is often terminated by the custodial entity through the use of force-feeding.

President Obama and first lady Michelle Obama made a surprise visit recently to speak with immigration advocates who were fasting for comprehensive immigration reform on the National Mall. The president and first lady thanked Eliseo Medina, Dae Joong Yoon and all of the fasters for their sacrifice and dedication, and told them that the country is behind them on immigration reform," said a White House official. "The president told them that it is not a question of whether immigration reform will pass, but how soon. He said that the only thing standing in the way is politics, and it is the commitment to change from advocates like these brave fasters that will help pressure the House to finally act.

CHAPTER EIGHT

THE BIBLE, OBESITY AND WEIGHT LOSS.

The Bible does not specifically address obesity and weight loss, however, there is much about the importance of taking care of our bodies and our health, and the commandments against gluttony that is over eating. The Old Testament, in Deuteronomy 14: 1-21, how God specifically warns and instructs about what His people, the Israelites, were to eat. Most of these were commands God gave them in order to keep them from eating harmful foods that would impact their overall health in a negative way. Some of the commands were given so they would not imitate the diet and habits of the idolatrous people around them.

Gluttony, which is act of overeating or drinking to excess, is mentioned in the Bible as being something to avoid (Prov. 23:20-21). Gluttony can cause health risks and become a drain on our finances. And the love of food and drink can all too easily become an idol in our lives. An idol is anything that takes the place of God or becomes our number-one focus, and thus a sin against God (Ex. 21:3-6).

Apostle Paul, in the New Testament, (1st Cor. 6:16-20) tells followers of Jesus Christ that our bodies are temples of the Holy Spirit, and as such we are to take care of the bodies and keep them as healthy as we can. And being obese leads to multiple health risks, we need to realize that as much as it is up to us in our choice and amount of food, drink, and exercise, we should strive to avoid becoming overweight.

The Lord wants His children to take excellent care of their bodies since they are the residence of the Holy Spirit. A strong, healthy body helps us stay in shape so we can better serve God each day and thus bring glory and honor to Him, our principal reason for living. The Lord wants us to keep our focus on Him and not fall into obsessing about weight gain, weight loss, or food and drink, any of which can become an idol in our lives.

Is gluttony/overeating a sin?

Gluttony seems to be a sin that Christians like to ignore. We are often quick to label smoking and drinking as sins, but for some reason gluttony is accepted or at least tolerated. Many of the arguments used against smoking and drinking, such as health and addiction, apply equally to overeating. Many believers would not even consider having a glass of wine or smoking a cigarette but have no qualms about eating themselves at the dinner table. This should not be.

Physical appetites are an analogy of our ability to control ourselves. If we are unable to control our eating habits, we are probably also unable to control other habits, such as those of the mind (lust, covetousness, anger) and unable to keep our mouths from gossip or strife. We are not to let our appetites control us, but we are to have control over our appetites. (Deut. 21:20), (Prov. 23:2), (2 Pet. 1:5-7), (2 Tim. 3: 1-9), and 2 Cor. 21:5) The ability to say "NO" to anything in excess self-control is one of the fruits of the Spirit common to all believers (Gal. 5:22).

God has blessed us by filling the earth with foods that are delicious, nutritious, and pleasurable. We should honor God's creation by enjoying these foods and by eating them in appropriate quantities. God calls us to control our appetites, rather than allowing them to control us.

Nine different types of fasting in the Bible.

While the purpose of fasting in the Bible has nothing to do with losing weight, a number of diets have come about through the reading of Scripture. It changed their attitude about food. It changed their lives and they reached healthy weights as a result. The same has happened when people have studied what the Bible teaches about fasting: For instance, that's where the Daniel Fast comes from, but even with the Daniel Fast, you'll find different versions based on different interpretations of the Scripture.

1.) The Daniel Fast

The Daniel fast idea came from the book of Daniel 1:8-14, in the Old Testament, where the king of Babylon wanted some certain children from Israel, without any blemish, well favored and had wisdom, knowledge and understanding, who had the ability to stand in the king's palace and be taught to speak in Chal-de-ans. These children Daniel, Shadrack, Meshach and Abednego were qualified to stand in the king's palace. They were asked to eat in the kings' palace. But they turn down the food and wine and requested only for vegetables and water for the period of ten days. During which Daniel and his friend were healthier than those who ate of the kings' diet. This goes to show how important vegetable and water are to the body. Eating right can be seen from Daniels story, is very important in our today's societies. The fat portion of pizza, hamburger, alcohol intake and smoking of cigarette, can later in life destroy the body. The best is to eat some food that will edify the body and bring them closer to God.

The Bible does not command God's people to follow the Daniel fast. This is an account of one young man's personal convictions regarding diet. He didn't want to be assimilated into the Babylonian culture. Turning away the king's food was one way to separate himself as one set apart for God.

2.) The Normal Fast

This is the regular fasting that is done from 12 midnights to 6 pm in the evening. During this period, you are praying to seek the face of God as to whatever problem they are facing for relieve. No food until 6 pm in the evening. This fasting last about 24 hours a day.

3.) Partial Fast

A partial fasting is usually called by the physician to persons trying to undergo surgery. They are only allowing to drink clear liquid juice or food like apple sauce. These are people with restricted diet. Daniel's vegetable fast is an example of a partial fast. Another example is to give up food between meals such as some people do for Lent. Since food is eaten, some people argue this isn't really a fast.

4.) Radical Fast

This is meant specifically for political reasons that a specific issues need to be addressed by the government. Another is an individual who wants to seek the face of God for an answer into an issue or matter bothering on a particular desire that needs to be addressed by God. These kind of fasting required neither food or water. For example, Jesus fasted 40 days and night in the wilderness without food or water. (Mathew: 1-11). The second form of radical fasting allows water but nothing else. Some other examples of fasting are Moses who went up to the mountain and fasted 40 days to seek the face of God (Deuteronomy 9:9-18). In 1st Samuel 1: 11, an account of Hannah who had no child and had to fast and pray for days and her request was granted. Others are Esther and her family (Esther 4: 15-16), and David in 2nd Samuel 12: 15-20.

5.) Fasting to Experience God's Power

For us to experience God's power and grace, we need to fast. By so doing we focus on God instead on ourselves. For example, Moses fasted to experience God's power and grace when he had to split the red sea into two parts. So did Esther who experienced God's power and grace. In Genesis 18: 13-14, Sarah was barren but she experienced God's power to bear a son at the age of 90. In Numbers 11: 21-22, Moses feed the Israelites in the wilderness with miraculous manna. And in verse 23 of Numbers 11, it is said that God's power is not limited. Nothing is too difficult for God. (Jeremiah 32:17).

6.) Corporate Fasting

Corporate fasting as a church or group of believers joins or comes together with a date set aside for a period of fasting and prayer. They are to fast and pray for a change in a particular problem or situation. During this period fasting will be done for God's Will to take place on that particular problem which they are fasting and praying for. An example of such was Esther and her family, when Esther told her uncle to fast and pray. Other corporate prayers are found in the book of Mathew 6:16-18, 1 Samuel 7: 5-6, Ezra 8: 21-23, Nehemiah 9: 1-3, Joel 2: 15-16, Acts 13: 1-3,14:21-23. This fasting can be done even on telephone conference calls. The fasting is most often connected with specific prayer requests. Samuel also fasted when he instructed the Israelites to separate themselves from the world. They had drifted far from God's will and were persecuted by the Philistines. Along with getting rid of the false gods, Samuel told them to fast and pray

7.) Samuel Fast

In 2nd Samuel 1:2, where Samuel and the people mourned, wept and fasted until evening, for Saul, and Jonathan his son, and for the people of the Lord, and for the house of Israel, because they fell by the sword. Samuel also fasted when he instructed the Israelites to separate themselves from the world. They had drifted far from God's will and were persecuted by the Philistines. Along with getting rid of the false gods, Samuel told them to fast and pray. Another of Samuel fast found in 1st Samuel 12:19, is where the people told Samuel to pray to the Lord on their behave, so that they will not die because they made their sins worse by asking for a king. And in 2nd Samuel 12:6 where David fasted and besought God for the son by Uriah, went in, and lay all night upon the earth. See other examples of Samuel kind of fasting in the Bible; Acts 9:9, Daniel 6:18, Nehemiah 13:31, Psalm 69:10 and Isaiah 22:12

8.) Paul's Fast

Paul's fasting helps us to benefit or gain insight and wisdom. (Acts 9:9). It is very useful when we are confronted with tough issues. Or when we are in the dark, confused and unclear about making choices, we then go into Paul's fasting. Paul was someone who persecuted Christ and His followers. And on his way to Damascus road, he was struck down by God, and was blind for three days. And during this period of his blindness, he neither ate food nor drank water for those three days. He was met with a decision that changed his life. With his blindness, he fasted and prayed as he seeks God's will for his life. (Acts 9 1-22). So if you are trying to make a decision about your weight lost, think about Paul's fating. When we decide to fast, God will cause His light to break forth like morning. (Isaiah 58:8). One way we can think of focusing to God for right choices and decisions is through fasting.

9.) John the Baptist Fast

In Luke 1 :15, "it is stated that "John shall be great in the Lord, and shall drink neither wine or strong drink". And in Luke 7:3, it is written that John the Baptist fast by setting himself apart for a time. He lived life in the wilderness and ate a limited diet of locusts and wild honey. While that isn't going to be the diet of choice for most people today, this fast represents separation from the world.

Because John the Baptist was the main messenger of Jesus Christ, he vows to fast to avoid wine and strong drink. That was his life style as he was set to carry on the special mission or message of Jesus Christ. We too today can set apart ourselves from unhealthy food choices that can cause harm to our bodies and to embraces a life of healthy life choices. The question is, when are you ready to eat right? Do you want to keep living your life eating unhealthy food?

CHAPTER NINE

RECIPES TO HELP CUT DOWN ON YOUR WEIGHT

Ripe and UN ripe Plantains

Photo by Sally Iyobebe

How to prepare unripe Plantain pottage or Porridge

Ingredients:

2 kg fresh fish, fresh tomatoes, or a tin of tomatoes paste, fresh or dry ground pepper, 2 medium size onion, Olive oil, 4 medium size tatashe (ground), 2 Knorr cube, Thyme 1 teaspoon, Curry leaves or spinach, water as desired.

1.) Cut the fish and wash thoroughly.
2.) Steam with few slices of onion and salt for about 10 minutes.
3.) Pour the olive oil into a pot.
4.) Add the ground ingredients (ground tomatoes, pepper, and onion) together with the tomato puree.
5.) Stir and fry for another 10 minutes.

6.) Add the Knorr cube and pour the seasoned fish.
7.) Wash and shred the curry leaves and add to the pot.
8.) Simmer for another 5 minutes.
9.) Stir and add salt to taste.
10.) Remove from heat.

Preparation of unripe plantain

Peel the unripe plantain

1.) Cut into four slices and wash
2.) Bring water to a boiling point
3.) Add the cut plantains to the boiling water.
4.) Boil for about 30 minutes, and make sure it is soft
5.) Add the already steam tomatoes
6.) Leave to cook for about 10 -15 minutes
7.) Taste for salt
8.) Remove and serve while still hot.

Baked ripe plantains

3 large, ripe plantains (skin totally blackened)

Preheat oven to 350°F. Wash and dry plantains. Trim off both ends of each plantain. Make a slit on the body of the plantain lengthwise. Place on a baking pan and bake for approximately 45 minutes (turning over halfway through the baking), until plantain flesh is tender. Slice each plantain into 3 equal-sized pieces.

Serve with vegetable soup, stew or with fried peanuts.

Unripe plantain can be dried and ground like flour for plantain fufu

It can also be boiled and eating with bitter leaf soup or ground nut soup (ndole), or stew

Unripe plantain just like the ripe plantain, can be fried and eating like snacks.

Efo Riro (spinach)

This vegetable is found all over the world. This rich vegetable soup is largely eating by the Yoruba's tribes of Nigeria. The vegetable that can be used to cook this soup are water leave or ugwu. It can also be cooked by itself.

Ingredients:

Spinach (Efo Riro).
Water leaf or Ugwu (optional).
Assorted beef (cow feet, shaki (beef stripe), chicken, oxtail, turkey).
Dried fish, stock fish, crayfish and Pomo,
Tomatoes, onions, thyme, garlic and ginger.
Palm oil (low cholesterol), pepper, salt and maggi and seasoning.

Method:

Soak the stock fish and dry fish for a few hours. Wash stock fish neatly and boil. The length of time depends on how hard the stock fish and adding small amounts of water at a time and top it up as you cook it. Continues until it becomes tender or soft.

For the dried fish, wash the fish, and remove the bones and separate them into small pieces.

1. Cut the frozen or fresh spinach, wash and rinse out the excess water fully.
2. Prepare other ingredients: grind pepper, tomatoes, cut the onions, ginger and garlic, and bring to boil into a paste.
3. Cook all the assorted beef with little water.
4. Add the stock fish and dry fish, seasoned with onions, salt and maggi.
5. Add some water and let it boil, then put in the crayfish, the paste r and palm oil
6. Cover the pot and cook till it is properly boiled and the oil has changed from red to yellow.

7. Add spinach (or any other soft vegetable, pumpkin, waterleaves) optional.
8. Cover again and leave to cook, taking care not to overcook the vegetables.
9. Taste for salt and leave to simmer again for about 15 minutes.
10. Serve with amala, fufu, pounded yam, rice, eba.

Bitter leaf soup

Bitter leaf soup is one of the most popular African traditional soups. It could be medicinal in the treatment of high blood or diabetes.

Ingredients:

Bitter leaves (washed and squeezed)
Cocoyam, red palm oil, assorted meat (cow feet, stripe, turkey chicken),
Assorted fish (stock fish, dried fish), Cray fish
Pepper, salt, maggi and seasoning to taste.

Method:

1. Wash and cook cocoyam until it get soft. Remove the peels and pound in a mortar into a smooth paste.
2. Wash bitter leaves and boil for about 15 minutes to remove some of the bitter taste.
3. Boil the Assorted meat, and the stock fish to be tender.
4. Add pepper, ground crayfish, salt and maggi and let it cook for about 10 minutes.
5. Then add the cocoyam paste (in small lumps) and the palm oil, and let it cook until the cocoyam is absorbed in the water. You can add more water if it too thick.
6. Cover the pot and leave to cook on low heat.
7. Finally taste for salt and the soup is ready.
8. Serve with eba, pounded yam and fufu.

Bitter leaf, Ugwu and Spinach Soup.

For ingredients and methods, see bitter leaf soup.

Egusi Soup

Egusi soup also known as melon soup is a delicious African soup that is made with ground melon seeds and enriched with assorted meat, fish and spices. It is one of the most popular African soups used for eating various meals such as pounded yams, eba, fufu and wheat meal. There are diverse cooking methods as you travel from place to place or country to country.

The common variations for cooking Egusi Soup include:

Egusi soup cooked with vegetables (spinach/ugwu, bitter leaves or a combination of both.)

Egusi soup cooked without vegetables

Boil the ground egusi seeds before adding the palm oil and condiments

Boiling the assorted meat and fish, then adding the ground egusi seeds, with or without vegetables. You could try out the different variations (if you can), and make your cooking methods your own preference. All that really matters is getting it right and making sure that the ground egusi seeds are properly cooked, to get the raw taste out.

Ingredients:

3 cups ground Egusi/melon seeds
Assorted meat of choice shaki, (beef stripe) beef, chicken, (cow's skin) pomo, cow feet
Assorted fish (stock fish, dried fish)
Spinach leaves/bitter leaf
1 medium sized onion bulb.

1/2 bottle palm oil or olive oil
Pepper, salt, fresh tomatoes, onions, maggi, and seasoning.

Method:

Wash and boil all the assorted meat. Wash and boil stock fish, dry fish. Wash your bitter leaves or ugwu. In a pot pour and heat, the palm oil, then turn in the grind egusi or melon and cook it in the stock and cover to boil for about 20 minutes. Leave to boil for 10 minutes. Add the already boiled assorted meat and stock fish, and leave to boil for 25 minutes. Then add your tomatoes, pepper. Onions, maggi, salt and seasoning, cray fish and cover it to boil. Stir the soup to avoid it stick in the pot. Add your bitter leaves or ugwu and allow to simmer for about ten minutes. Taste for salt and maggi and serve while it is still hot.

Egusi Soup can also be eating with boiled rice but it is mostly enjoyed with semolina, amala, tuwo masara, fufu, pounded yam, and several other bolus meals.

Edikang Ikong Soup:

The Nigerian edikang Ikong soup or simply vegetable soup is native to the Efiks, people from Akwa Ibom, Cross river States of Nigeria and the Cameroons. Edikang ikong soup is very nutritious and the most expensive among all the vegetables soups. In most cases because of its nutritional value and the cost, it is mostly eating by the rich. It is prepared with generous quantity of ugwu leaves and water leaves, the soup recipe is nourishing in every sense of the word. This is a rich African vegetable soup, commonly served as a number one delicacy during important occasions.

Ingredients:

1. Assorted meats (cow feet, oxtail, Shaki (beef stripe), Pomo, and bush meat).
2. Stock fish, dry fish, cray fish, periwinkles, snails.

3. Fresh ugwu, (shredded) water leaf (washed and if possible shredded).
4. Salt, pepper, Palm oil (low calories/cholesterol), maggi, onions etc.
5. Where you cannot find water leaves, use fresh or frozen spinach or lettuce.

Method:

Wash and cook the assorted meat.

Add salt, onions and maggi and some seasoning to the assorted meat.

Leave to cook for about 45 minutes

Wash snail with lime, alum to remove all slime from the skin.

Wash stock fish and boil, wash dry fish with salt to remove all the dirt and sand. Rinse with warm water.

Wash the oxtail and cook in the pot of the assorted meat and boil again for another 20 minutes.

Add the water leave and allow to boil for 15 minutes and then add the ugwu and, periwinkle, mix properly.

Allow to cook for 20 minutes, and then add the crayfish, pepper, salt and palm oil.

Leave to cook again for 10 minutes, then mix and stir gentle and cook for another 10 minutes.

Serve with pounded yam, fufu, semolina and eba.

Eru or Afang Soup

This is another tasty dish and more expensive than the Edikang-Ikong from the Cross River, Akwa Ibom State, the Ibos and the Cameroons. Consisting of basically vegetables, this soup is rich in Vitamins, Protein and fiber. It is mostly prepared for ceremonial occasions like weddings, important personnel etc. This is because the ingredients used in the preparation are cost effective. It is mainly for those who have the money that can afford to cook this particular soups.

Ingredients:

Afang leaves or Ukazi, water leaves or spinach.
Assorted meats (cow feet, oxtail, stripe, pomo, and bush meat)
Stock fish, dry fish cray fish, periwinkles, snails
Salt, pepper, Palm oil (low calories/cholesterol), maggi, onions etc.
Where you cannot find water leaves, use fresh or frozen spinach or lettuce.

Method:

Wash and cook assorted meat until tender.

Wash and cook stock fish until soft. Add water leaves or spinach and cook for about 5 minutes

Add Afang leaves (pounded or thinly sliced). Then add crayfish, palm oil, salt, pepper and maggi and some spices like ginger or garlic. Cover pot and allow it to cook for about 20 minutes. Make sure not to overcook the vegetables. Taste for salt. Serve with eba, semolina, semovita, pounded yam, cassava fufu.

Otong Soup

Otong Soup is a very rich Okro and vegetable Soup from the Cross River area (Calabar)

Ingredients:

1. Assorted meat(cow feet, oxtail, Shaki,(beef stripe), pomo, bush meat, hen and turkey)
2. Stock fish, dry fish, periwinkles, dry prawns, crayfish,
3. Okra, ugwu/pumpkin leaves, onions, pepper, salt, maggi, ginger, garlic and seasoning.

Method:

Wash the meat and cook in a big pot. Add salt, maggi, onions, garlic and ginger

Wash the dry fish with salt and warm water to remove sand from the skin.

Add the stock fish, dry fish, dry prawns and periwinkles to the cooked meat and leave to cook for 15 minutes. Add the shredded ugwu/pumpkin leaves to cook for 10 minutes. Add okra, crayfish, olive oil, pepper, maggi and salt to taste.

Stir the soup and allow it to boil until properly blended. Remove from heat and serve with cassava fufu, pounded yam and eba

Ekwang (Ekpang Nkukwo)

Ekwang is known as ekpang nkukwo is a delicacy from the Efiks and Ibibio's. It is also widely enjoyed in Cameroon and very popular with the Bakweri tribe in the southwest part of the country. It can be used for any occasion but mainly for the royalties. It is cost effective and time consuming.

Ingredients:

Cocoyam (as desired)
Cocoyam leaves, spinach or bitter leaves
Assorted boiled meat (beef). Dried fish, ground crayfish
Onion chopped, pepper, salt, maggi, red palm oil and grater

Method:

Cut and wash all the assorted beef. Season and boil until soft.

Peel the coco yam, wash and grater until it become puree.

Wash the cocoyam, spinach or bitter leaves, making sure all the sands are removed.

Put in small amount of salt in the grated cocoyam and mix until the salt is absolved in the cocoyam.

With a teaspoon, scoop the grated cocoyam and place it in the leaf and wrap it firmly.

Put in a pot of oil round it until all the process is complete.

Boil water and pour it into the wrap cocoyam and place pot on the stove. Leave to boil and then pour in the stock, pepper, cray fish and fried fish, salt and maggi and leave to cook for 20 minutes. Cook in a low heat.

Do not stir because it will become mushy. Just bring the pot out from the stove, and with both hands, shack or move the pot around to prevent it from burning under.

Leave to boil for another 15 minutes and then shack the both. Do this continuously until the Ekwang is properly cooked.

Taste for salt and pepper. Remove and serve hot.

Banga Soup:

Banga soup is made from fresh palm nut fruit. It is boiled and pounded in a mortar and with hot water wash and squeeze the juice from it. Use a sieve to make sure the chaffs do not go into the juice. That is separating the kernels from the chaff. This soup is unique in the sense that it is completely of its natural ingredients. Fresh palm oil is usually recommended for stew rather than the usual groundnut or vegetable oil. This soup is popular with Edo and Ibo people and some tribe from West Cameroon.

Ingredients:

Banga (palm fruit), Banga paste for those abroad
Assorted Meat (cow feet, oxtail, Shaki, (beef stripe), turkey. Hen or chicken), or fresh fish or fresh shrimp
Stock fish, dry fish, crayfish, bush meat
Onions, pepper, salt, maggi, and spices (ginger, garlic scent leaves).
Achi or okro as your thickener
Bitter leave or ugwu

Method:

Wash meat and place in a pot, add water as needed and season with salt, onion and maggi.

Add the already washed stock fish and cook for another 20 minutes,

Extract the oil from the already boiled palm nut.

Pour the strained oil or pulp into the already cooked meat, and add pepper, salt, maggi, onions and all the available seasoning.

Add achi or okro to the soup to thicken it. Add crayfish and leave to cook for about 20 minutes

Taste for salt. Serve with Starch, pounded yam, fufu, and eba.

Gbegiri (beans soup)

This is another nutritious soup by the Yoruba's of Nigeria. It is use to mix with ewedu when eating with amala or fufu. It can also be eating with boiled rice or by itself as it serves as a repairer to a dysfunctional body. Those who do not like to eat meat, this will be the best soup for them.

Ingredients:

Brown beans, black eye peas or white beans is recommended.

Olive Oil or palm oil stock (Dry) Fish, Assorted beef, salt, Maggi, Grounded Crayfish, Dry Pepper or Blended pepper and Tomato and spinach

Recipe

1. Soak beans in water and remove the back of the beans
2. Boil the assorted beef until tender.
3. Boil the beans until soft, leave to cool and then blend it in a puree form.
4. With a sieve, extract the liquefied beans in a pot and boil.
5. Add your olive oil or palm oil, pepper, salt and maggi.
6. Boil for another 10 minutes before adding all the assorted meat, crayfish, and dried fish.
7. Leave to cook for another 10 minutes.
8. Taste for salt. Remove and serve with amala or pounded yam.

Allow to boil for 3-5 minutes, then add the meat stock. Season to taste, Add the dry fish (ensure you soak the dry fish in some hot salted water wash before using it) and the Cray fish. Allow to boil on low heat for 10 minutes.

You can serve with ewedu (jute leaves) and pounded yam or amala. It is best serve with Amala.

Moin-Moin

This could be used with black-eyed beans, brown beans or white beans.

Ingredients:

Beans, tomatoes, onions, fresh pepper. Ground crayfish, salt, maggi olive oil, water of beef stock

Methods:

Soak and wash beans, robe to remove the peels. Or you might want to buy the already peeled beans which you just soaked it overnight to make it soft.

Put the beans in a blender and grind. When done, also blend the pepper, tomatoes, onions, and ginger.

Add the blended pepper, tomatoes, onions, ginger, maggi and salt into the beans paste, including the crayfish.

Mix until all the ingredients are absorbed into the beans paste, and it becomes a smooth paste.

Add a little bit of warm water to loosen the paste but remember not to make it too light.

Using an empty milk or tomato container or aluminum foil or sandwich bags and pour in the paste to a level that it will not overflow, and have enough room to expand once it start boiling or cooking.

In a big pot, place the containers of the paste with little water occasionally, cover and let it cook for 45 minutes and the results will be that of a baked care.

Then serve with rice, pap, garri, boil or baked or fried plantain.

Beans, yam or water yam

Beans, yam or water yam

Dried fish, crayfish, dry pepper, onions, olive oil, salt, maggi, ginger, garlic, spinach and seasoning.

Method:

Wash beans and boil until soft.

Peel and wash yam cut into smaller pieces and add to the beans stir and leave to boil

Add the dry fish, cray fish, pepper, salt, maggi, olive oil and seasoning.

Cook until the yam is soft, stir and add salt and maggi to taste.t

Add spinach and leave to cook for 5 minutes.

Serve hot.

Beans pottage

Ingredients:

Beans, dried fish, crayfish, dry pepper, onions, olive oil, salt, maggi, ginger, garlic and spinach, seasoning.

Method:

Wash beans and boil until soft.

Add the dry fish, cray fish, pepper, salt, maggi, olive oil and seasoning.

Cook until all the ingredients have blended together, stir and taste for salt.

Add spinach and leave for 5 minutes.

Serve hot with rice, yam, sweet potatoes, fried or boil plantain.

Jollof Rice

Most popular African dish prepared at parties.

Ingredients:

Rice, Chicken, tomatoes, fresh pepper, onions, mixed vegetable, salt, maggi, thyme, curry and seasonings., olive oil, stock.

Before you cook jollof rice make sure the stew is ready. You can always keep the stew in your freezer so that it does not consume too much of your time when you are ready to cook.

1. Parboil the rice if need be for about 20 minutes to remove excess starch.
2. Place in a filter to drain the water from the rice.
3. With a large pot, pour in the already prepared stew and leave to simmer for about ten minutes.
4. Then pour in your stock and bring it to boil for about 20 minutes.
5. Pour in the rice, add maggi, salt and seasoning to taste. Reduce heat and leave to cook slowly.
6. Keep checking on the rice. Where it is still hard, add little water until it is well cooked. Add your mixed vegetable and let it cook for just 5 minutes.

Serve with your chicken and vegetable salad.

Fried Rice

Another popular dish use for parties. Very delicious when prepared very well

Ingredients:

Rice, mixed vegetables, green pepper, thyme, liver, shrimp or chicken, onion, garlic, curry, nutmeg, olive oil, seasoning, salt, magic and butter.

Method:

Cut and wash the chicken, or liver or shrimp. 2. In a pot, boil the chicken or liver until tender.3. Wash and parboil the rice, drain the water from it.4. with the chicken out from the pot, and depending on the kind of rice, pour in the rice in the chicken stock, together with all the green pepper, mixed vegetables and butter. Taste for salt and cook on low heat until soft and water dried.

Then serve hot with some salad.

Coconut Rice

Ingredients:

Coconut milk, rice, ground chicken or turkey, dried fish, crayfish, pepper, onions, salt, maggi, ginger, garlic.
Fresh vegetables. (Carrots, peas, beans, green pepper). Optional

Method:

Peel, slice and blend the coconut into a smooth paste.

Wash with warm water to remove the coconut milk.

In a pot, boil the coconut milk; add the dried fish and assorted meat.

Add the par boil rice, crayfish, onions, pepper, salt, maggi and sliced tomatoes

Cover the pot on low heat allows to cook for 30 minutes.

Add mixed, green pepper as garnishment(optional), then serve.

Ogbono Soup

Also known as draw soup, makes swallowing of food very easy due to its slimy nature as it help the food go down the throat smoothly without any hindrances. Depending on what you want, you might want to cook it plainly or with vegetable. (bitter leaf). But for the purpose of this book, I will encourage that no matter the meal you cook, it should have some kind of vegetables in it. You may want to add either okra or egusi or bitter in the ogbono soup.

Ingredients:

Ogbono seed.
Assorted beef: Cow or beef feet, shaki (beef stripe), chicken, turkey.
Dried fish, stock fish, cray fish.
Palm oil-low calories.
Vegetable: fresh spinach or bitter leaf.
Pepper and salt to taste.

Method:

Grind the ogbono into a paste or powder form.

Cut the spinach and wash clean to remove sand.

Wash bitter leaf to extract some of the bitter taste from it.

Cook the assorted beef, the stock fish until tender or soft.

Grind crayfish and pepper.

Have your boiling water ready.

Directions:

First pour palm oil in a pot at a low heat, and turn in the already grind ogbono.

Stir with a wooding or cooking spoon to let it dissolve in the oil

Add the stock from the meat and let it cook for about 20 minutes to make sure the ogbono is completely dissolve and thickened.

Add the hot water to make sure it has absorbed all the water. Then pour in your assorted beef and stock fish and leave it to cook.

Keep stirring to avoid it sticking at the bottom or base of the pot.

Cover the pot and leave to cook for about 5 to 10 minutes and stir again.

Pour in the stock fish, dried fish and crayfish to add to its flavor

Add salt and magi to taste.

Add your bitter leaf or spinach which ever you prefer.

Serve with Amala, eba, pounded yam or fufu.

Okra Soup (Miyan Kubewa)

Ingredients:

Okra.
Assorted beef: cow feet, stripe, chicken, turkey.
Fresh fish, stock fish, crayfish.

Salt, pepper, onions, palm oil (Low calories,), maggi, bitter leaf or ugwu, and potassium. (Akuan).

Method:

1. Wash, slice or grate okra into tiny pieces.
2. Cut and boil the assorted beef, add salt, maggi and onions to taste.
3. Wash and boil stock fish, add salt, onions and maggi

4. Grind crayfish into a powder form.
5. Pour the stock water into a pot and boil. Then pour in the already cut okra and add your potash to help make the okra slimy.
6. Allow to cook for 15 to 20 minutes to allow it absorbed in the water.
7. Then put in your oil, assorted beef and stock fish and allow it boil for ten minutes.
8. Add your dried fish and crayfish and allow it to cook for another 5 minutes
9. Add salt and maggi to taste.
10. Serve with tuwo chincapa, Amala, eba, pounded yam Cassava fufu.

Ewedu Soup

Ewedu soup is indigenous to the people of Yoruba, a very popular Nigerian tribe. Having leaved in Lagos for 13 years, and School at the University of Lagos I can proudly talk about Yoruba food because I have had my own share. Ewedu soup is usually served with stew gbegiri.

The ingredients:

Ewedu leaves
1 teaspoon powdered potash
about one to two cups of water
Ewedu Broom (local broom used in mashing it)
Salt and maggi to taste

There are more simple ways to prepare this soup without the use of the mashing broom. You can slice the leaves to tiny bits and commence cooking or you can even blend it.

One of the qualities of the ewedu leaf is the ability to draw, perhaps the reason you cannot make this soup with any other leaf.

Method:

1. Wash the leaves thoroughly with clean water to remove sand.

2. pour about two cups of water to a cooking pot and heat to boiling point, add the washed ewedu leaves, the potash (to soften the leaves), cook for about ten minutes.
3. Then use the cooking broom to mash (more like pound) continuously inside the pot, this will turn the leaves to tiny bits after mashing for about five minutes.
4. Add salt, maggi and pepper to taste.
5. Serve ewedu soup plus assorted stew with Amala, eba, semolina or pounded yam, the exact way a Yoruba man would love it.

Miyan Yakwa:

Ingredients:

Cow feet bones, smoked or dried fish, and Cray fish.
Groundnut paste, fresh pepper, tomatoes, onions, olive oil, salt, pepper, maggi, dawadawa, yakwa leaves,

Method:

Wash the soft bones and season with onions, pepper, maggi and salt.

Ground the tomatoes, dawadawa, and tatashe into a smooth paste.

In a hot pot of olive oil, add the ground ingredients and fry for 15 minutes

Add the fish and soft bones with the stock and stir, add water, maggi.

Wash the yakwa leaves, cut and add to the soup.

Cook on low heat and add the ground nut paste.

Allow to cook for 15 minutes. Stir and taste for salt.

Remove and serve with tuwo dawa.

Oha Soup

This soup is similar to bitter leaf soup but is cooked with Ora leaves

Ingredients:

Ora leaves, cocoyam, palm oil (low cholesterol), Assorted beef (tripe, cow feet, oxtail, pomo),
Stock fish, dry fish, pepper, salt, and cray fish.

Method:

Wash and boil cocoyam until soft, remove peels and pound in a mortar until smooth paste.

Use your fingers to cut the oha leaves into small pieces.

Boil the assorted meat, stock fish and dry fish

Add salt, maggi, cover and leave to cook for 10 minutes

Add pepper; ground Cray fish including the cocoyam paste in small lumps and palm oil.

Cover the pot and leave to cook on medium heat until all the lumps have been dissolved or absorbed in the stock.

Add the ora (oha) leaves and leave to cook for 10 minutes.

Taste for salt and add some water if it is too thick. Serve with eba, fufu, and pounded yam.

Groundnut Soup

Ingredients

Assorted Beef (cow feet, oxtail, Shaki (beef stripe), Pomo, turkey, hen or chicken)
Pepper, onions, salt, maggi, seasoning, groundnut paste, ginger and garlic, bitter leaf

Method:

Cook assorted beef /hen, chicken with onions and salt

Leave to cook until tender

Add pepper tomato puree and leave to cook for 15 minutes

Add the peanut butter and then cook over a gentle flame until the oil from the peanut butter starts coming to the top.

Put the already cooked assorted meat into the ground nut paste and leave to cook for 20 minutes.

Cook on a low heat.

Continue to add water intermittently as not to let it stick to the bottom of the pot and for it to become too thick.

Continued stirring gently with a wooden spoon over a gentle flame, until the peanut butter soup is ready for eating and is guaranteed to taste delicious.

Add a little bit of bitter leaf to it. (Optional)

Add maggi, Knorr cube, salt and seasoning to taste

Serve with pounded yam, Eba (Garri), rice, semolina

Pumpkin Leaves Recipe:

Ingredients:

Pumpkin leaves, Tomatoes, Onions, Pepper, salt, maggi, hen
Cow feet, shaki (Beef stripes), Palm oil, Ginger and garlic, Cray fish
Dried fish

Method:

Wash the pumpkin leaves to remove any sand from it

Chop the leaves thinly into fine pieces

Wash and boil already chopped hen, cow feet and beef stripes

Blend the tomatoes, pepper, garlic, ginger and onions together

Heat the palm oil in a pot and turn in the already blended tomatoes, pepper etc. and leave it to cook for 15 minutes. Make sure it does not have any sour taste.

Pour in the already cooked hen, cow feet and beef stripes and leave to cook for another 20 minutes.

Pour in the washed dried fish, stir and leave to simmer for 10 minutes.

Add the sliced pumpkin leaves and leave for another 5 minutes. Add, cray fish, maggi and salt to taste.

Serve hot with either boiled rice, pounded yam, eba (Garri), or semolina.

Sally Iyobebe

Curry Chicken

Ingredients:

I packet of chicken.
Salt, Black pepper, Curry powder, Corn flour, Garlic blended.
Red and green bell pepper, finely chopped, 2 Large cooking onion, thinly chopped.
Olive oil for cooking, Mix vegetables, Fresh or dried thyme, Fresh tomatoes, Water to cook.

Method:

Wash chicken with warm water to remove any fat from it

Season chicken with salt, black pepper, curry powder and chopped garlic and ginger;

Add cooking oil to saucepan and heat on high, then add chopped onion, green and red bell peppers.

Add thyme, pepper sauce, curry powder, salt and tomato; sauté onion and peppers until just tender.

Add about 2cups of boiling water and, when sauce is boiling, add chicken. And leave to cook for about 5 minutes. Add seasonings; and another 2 cups of boiling water; cover and bring to a rolling boil. Finally add corn flour to thicken the chicken curry soup

Lower heat to medium, add the mix vegetables and corn flour to thicken the sauce cover and simmer (slow boil) until chicken is tender and sauce thickens for about 20 minutes.

Taste for salt. Serve with rice.

Fish, chicken/hen/turkey or assorted beef or meat stew

Ingredients:

Fish, chicken/hen/turkey or assorted beef or meat stew, fresh tomatoes, onions, clove, garlic, tomatoes puree, thyme, curry, garlic, ginger, fresh pepper, parsley or scent leaf and olive oil.

Salt, Maggi, Knorr cube and seasoning

Method:

Wash and cut Fish, chicken/hen/turkey or assorted beef or meat into pieces and season with salt maggi, thyme and sliced onions and curry and cook for about 25 minutes.

The fish is not to be cooked or boiled

Grind tomatoes, fresh pepper, garlic, ginger, onions etc.

Fried the already grind tomatoes, fresh pepper, tatashe, garlic, ginger and onions in olive oil in a pot until the water is dried and in a paste form, making sure it has no sour taste

Add the fish or chicken/hen/turkey in the paste with some water or stuck.

Leave to cook for about 20 minutes

Add salt, maggi, Knorr cube, curry powder, scent or parsley and seasoning

Taste for salt again to make sure it is tasty and delicious

Serve with rice and beans, boiled ripe or UN ripe plantains, fried ripe plantain, boiled yams or potatoes, cuscus etc.

Cabbage Stew

Ingredients:

1 large cabbage.

Chicken, Cray fish, stock fish, cow feet and tripe or assorted meat. Fresh tomatoes, tin tomatoes and onions.

Salt, maggi, pepper, olive oil.

Method:

Wash and boil cow feet, tripe stock fish and chicken.

Wash and sliced cabbage thinly and boil until soft.

Blend tomatoes, pepper, onions, ginger, garlic and cook in olive oil.

Boil cabbage and drain and quizzed water from cabbage, making sure there is none left.

Add the already cooked chicken, cow feet, stock fish stripe into the pot of tomatoes and other. Ingredients. Stir, cover and leave to simmer for about 20 minutes.

Add the cabbage, salt, maggi and other seasoning. Cook for 15 minutes.

Taste for salt. Serve with boil rice; boil green plantain, sweet potatoes etc.

Stuffed green or red bell pepper

I packet ground turkey.
6 green or red bell pepper.
3 large tomatoes, Garlic, ginger, salt and maggi or seasoning.
Olive oil, onion and fresh pepper.

Method:

Preheat oven to 300-350 degrees F.

With a low heat, Cook the ground turkey until brown and tender.

Remove and discard the tops, seeds, and the inside of the bell peppers. Arrange peppers in a baking dish with the hollowed sides facing upward. (Slice the bottoms of the peppers if necessary so that they will stand upright.)

In a bowl, mix the browned turkey, tomato sauce, garlic, onion all blended, salt, and pepper and olive oil. Spoon an equal amount of the mixture into each hollowed pepper. Mix the remaining tomato sauce and seasoning in a bowl, and pour over the stuffed peppers.

Bake for 1 hour in the preheated oven, basting with sauce every 15 minutes, until the peppers are tender.

Taste for salt.

Serve with boil rice, baked sweet potatoes or yams.

Fried Yams in butter

Ingredients;

Yams, butter, eggs, olive oil, salt and maggi to taste

Methods: peel and boil yams with little salt added to it.

Mix eggs in a bowl and add salt and maggi

Pour olive oil in a pot and pre-heat it

With the mixture of eggs in a bowl, put in the boil yam so that it absolves in the egg mixture

Remove and fry lightly until brown.

Serve with vegetables sauce.

White Soup (Nsala Soup)

Ingredients:

Assorted meat (pomo, cow feet, oxtail, hen, beef, fresh fish).

Cray fish, dried fish, cocoyam or yam as thickener, utazi, uziza, pepper, salt maggi.

Method.

Wash and cook the assorted meat until soft.

Boil yam or cocoyam and pound

In another pot, pour in the stock from the assorted meat, then add the pounded yam or cocoyam, pepper, salt, and maggi and leave to cook for about 30 minutes.

Continues to stir until the pounded yam or cocoyam is completely dissolve and smoothen.

Then add the assorted meat, cray fish, and dried fish and let it cook for 10 minutes

Add uziza or utazi. Taste for salt.

Serve with fufu, eba, or semovita.

Please bear in mind that as you prepare your soup to eat with your pounded yam, eba, semovita, cassava fufu, starch etc., be advise that just a little quantity of it is advisable if you want to cut down or maintain your weight.

CHAPTER TEN

SNACKS

Plantain Chips

Ingredients:

Green plantains, olive oil for frying, hot pepper or garlic, cloves for flavor.

Method:

Peel the green plantains or ripe plantains. Slice the green or ripe plantains in a cube form, full length, square or whatever shape you want. Mix or sprinkle with some salt before frying.

Heat the olive oil, in a deep frying pan, with enough oil to cover the plantains. Add the green or ripe plantains into the oil, fry until the chips get a brown or golden color. Make sure you put in just enough quantity of the plantain to avoid stickiness.

Fish pie

Ingredients:

Fresh Fish, potatoes, onions, carrots, garlic, thyme, curry, ginger, cloves, black pepper, olive oil and tomatoes.

Method:

Stem fish remove bones and cut or chopped into thinning pieces. Crumble it in a pot, add garlic, onions, thyme, cloves, salt, black pepper, maggi. Stir until it is cooked and brown.

Peel potatoes and cubed it and boil for about 5 minutes, drain the water and mash it.

Mix the already cooked fish and potatoes into a large pot and stir properly.

With the already mixed pastry of flour with butter, salt, baking powder and water, place on a clean table top or counter top. Roll with a wooden roller until smooth. With a desired flat 6 inches plate, place it on the smooth dough and cut round with a table knife. Fill the fish mixture, and place in the middle of the dough, cover it, with a fork in a mix bowl of eggs, cover and smoothing the edges. Continue until the whole exercise is completed.

Preheat the oven to 220 to 250 degrees C. with a big aluminum pan, grime the bottom of the pan to avoid any stickiness, brush the fish pie tops with milk or egg. Place in the oven and baked for 20 minutes, flip the other side and continue to baker until it is brown.

Serve with a bottle of orange juice, coffee or tea or water.

This method can be used for meat pie as well.

Fish Roll

Ingredients:

Fresh Fish, potatoes, onions, carrots, garlic, thyme, curry, ginger, cloves, black pepper, olive oil and tomatoes.

Method:

Combine the fish with bones removed, onion, carrot, egg, parsley, pepper maggi, thyme in a large bowl. Mix until it is completely absorbed.

Roll out the pastry and cut into a desired lengths, with a spoon full of the fish mixture, place at the center or middle of the dough and cover. Repeat with the remaining mixture and pastry.

Preheat the oven to 220 to 250 degrees C. with a big aluminum pan, grime the bottom of the pan to avoid any stickiness, brush the fish roll top with milk or eggs. Place in the oven and baked for 20 minutes, flip the other side and continue to baker until it is brown.

Serve with juice of any form, tea, coffee or water.

This method can be used for sausage rolls as well.

Puff-Puff

Ingredients:

Flour, yeast, ground nutmeg, brown sugar, salt, lukewarm water for mixing, olive oil for frying.

Method:

Put flour in a bowl, add the ground nutmeg, powdered yeast, brown sugar and salt. Mix all together by adding lukewarm water in small quantity at a time until you have a good and smooth blend. Use your hands when mixing the puff- puff ingredients or any other pastry. The mixture of warm water allows the dough to rise faster. Keep mixing the batter until it is very smooth and thicker than pancake batter.

Cover the bowl and leave to rise over night or for a couple of hours. Then fry with olive oil.

Chin-Chin

Ingredients:

Flour, brown sugar, baking powder, eggs, nutmeg, margarine or butter, salt, warm water and olive oil.

Method:

Mix flour together with baking, powder and salt into a bowl. Mix with butter, add eggs, sugar, nutmeg and water. Mix into a smooth stiff dough. Turn into a floured table and roll out thinly. Cut into thinning stripe or pieces and cut into small squares. Then heat olive oil in a deep frying pot and fry the dough in small batches turning frequently until it gets brown or golden color. Drain and allow to cool and then serve. Chin-chin can be store for several weeks or month in the refrigerator or an airtight container.

Doughnuts

Ingredients:

Flour, brown or icing sugar, baking powder, salt eggs yellow part, milk olive oil, cinnamon powder and oil for frying.

Method:

Put all your ingredients in a large bowl. Knead them into dough. Make them in a round shape. Using the doughnut cutter, cut them into a round shape while using a rolling pin. Fry them in a deep pot of oil until you have a brown color. Remove and sprinkle some brown or icing sugar on it.

Sally Iyobebe

Akara beans

Ingredients:

Beans, onions, fresh pepper as desire, salt, olive oil, warm water for mixing.

Method:

Soak already peel beans overnight to allow it get soft. Wash and grind until smooth.

Place in a large bowl season with salt, maggi and pepper with some warm water, small quantity at a time until you get the desired thickness. Sprinkle in the thinly slice onions and mix again with a metal spoon. With the already heated oil, scoop the beans mixture with a metal spoon and fry until brown or golden color. Drain the oil into a sieve and serve with pap, bread, tea or on its own.

Fresh Fish pepper Soup
(Chicken, Turkey, Oxtail, Shaki (Beef stripe))

Ingredients:

Fresh fish, saint leaf, maggi, garlic, ginger, onions, pepper, and pepper soup seasoning, bitter leaf.

Wash fish with clean warm water and keep in separate bowl.

Grind pepper, onions, garlic, ginger, and pour in a pot of water and bring to boil. Add the maggi, salt and leave to boil for another 10 minutes. Then add your already wash and season fresh fish in the pot of boiling water and cook for 15 minutes.

Please note, do not stir the fish because when you do it will scatter in the pot. Just hold the two handle of the pot, bring it out from the heat and shake it very well and gently, and place it back on the heated stove to cook for another 5 minutes.

Then remove and serve with boil rice, boil plantain, boil yam, baked or boil sweet potatoes etc.

This method /ingredients can be used to cook chicken pepper soup, stripe or assorted beef pepper soup.

Goat meat pepper soup
Ram meat pepper soup
Oxtail pepper soup
Assorted meat pepper soup
Chicken pepper soup
Catfish pepper soup.
Shaki (beef stripe) or cow feet pepper soup
All the above can be eating with agidi, boil yam, rice or sweet potatoes.

Dodo Oniyeri

Ingredients:

Plantains, eggs, salt and olive oil

Method:

Wash peel and cut the plantain into thin diagonal slice. Break the eggs into a clean bowl, season with salt and whisk lightly. Heat the oil in a pan, dip the sliced plantain in the egg and shallow fly gently on both sides until brown. Drain on kitchen paper and serve hot.

Kuli-Kuli

This snack is found mostly in the Northern Nigeria.

Ingredients:

Fried or Roasted Groundnut, Ginger, pepper, salt or maggi to taste, groundnut oil, onions, cloves, and flour.

Method:

Grind the roasted or fried ground nut either with a blender, mortar or grinding stone

Make sure it is soft and smooth

Add the already ground powder ginger and some ground pepper

Add salt and maggi to taste.

Mix until the ginger, pepper, salt and maggi is embedded into the groundnut.

Using a clean white cloth put in the mixed groundnut and squeeze until all the desired oil is out.

Then place the groundnut paste in a bowl and mold the Kuli-Kuli to your desired shape or size.

It could either be molded in a round, square or long shape.

Making sure you put some flour in your palm to avoid it from sticking into the palm.

With a deep frying pan, pour in your groundnut and allow the groundnut to heat before frying.

Fry until a golden or brown color is seen.

If you do not like to fry, you can use the same process and bake in the oven.

When completed Kuli- Kuli becomes crunchy when you eat it, except if it is very soft.

You can use it for a wedding, birthday, or any ceremony. It is a very healthy snack.

Peppered Gizzards:

Pepper gizzard is a snack that is good during the winter or raining season as it can be used as a heat warmer to the body.

Ingredients:

4 packs of chicken gizzards or more, lemons, lime, grind red pepper, curry power, seasoning, maggi cube, garlic, cloves, ginger, red bell pepper, tomatoes, red onion, white wine vinegar, wooden spoon and olive oil.

Method:

Wash gizzards thoroughly to remove all the impurities from it.

Then boil the gizzards until soft.

Grilled the gizzards in an oven for about 15 minutes.

Then blend the pepper, onions, garlic, ginger, few tomatoes, cloves, alligator pepper.

Bring to boil until all the water is dried up. Add some olive and fry a little.

Pour the gizzards in a clean large bowl.

Add the already fried pepper sauce into the bowl of gizzards.

With a wooden spoon, mix all together.

Serve with a cold glass of orange juice.

Peppered Snail

Peppered goat meat

Peppered chicken

Peppered dried or fresh fish.

Peppered ram meat.

Suya.

Vegetable soup

Ingredients:

Celery, carrots, onions zucchini and salt.

Method:

Wash and blend all the above vegetables into a liquid.

Then pour liquid or the juice into a hot pan.

Stir and mix properly and serve.

Cream of tomatoes soup

Ingredients:

Corn flour, Tomatoes.
Olive oil, Salt and maggi to taste.

Method:

Wash tomatoes and blend into juice.

Then pour juice into a hot pot or pan.

Add some corn flower to thicken it.

Add a tea spoon of olive oil.

Add salt and Maggi to taste.

Broccoli Salad:

Ingredients:

I bundle of broccoli, Vinegar, Chicken or shrimps or lobsters.
Bowl, onions, sugar, salt and seasoning to taste.

Method:

Place broccoli, onion, and cooked and shredded chicken in a salad bowl, and set it aside. In another small bowl, combine sugar, mayonnaise and vinegar; together, mix until the sugar is dissolved and the mixture is smooth.

Just before serving, pour dressing over salad in the bowl and mix.

Serve with whole grain wheat bread.

Macaroni Salad

Ingredients:

I packet of uncooked macaroni.
1 dozen uncooked eggs, onions, relish. Black pepper, yellow mustard.
Salt, mayonnaise olive oil.

Method:

With a pot of lightly salted water, bring it to boil. Then place the macaroni in the boiling water and leave to cook for about 15 minutes on low heat. Drain water away.

Boil your eggs to hard boil and grated the eggs.

Mix the macaroni in a bowl, grated eggs, slice onions, relish olives and a little olive oil.

Add mustard and mayonnaise as desired or needed. Add salt, black pepper or any season to taste.

Garnished with tomatoes, sliced eggs and organic spinach.

Then serve.

Isi Ewu

Ingredients.

Goat head, fresh utazi salt, maggi, pepper, nutmeg, onions, palm oil, ugba (oil bean seed), potash (Akaun).

Clean goat head by burn it to remove all the hairs from the goat head.

Open the mouth of the goat head and scrape the tongue and remove all the dirt from the tongue.

With an iron sponge, scrape the head of the goat to remove the smoky odor and burnt from the goat.

Put in a pot and cook until tender and soft.

Remove and cut into smaller pieces.

In a different pot, put in the akaun, palm oil until it is thicken, and then add all the other ingredients.

Add the cut pieces of goat head into the mixture, mix to make sure the ingredients have absolved into the goat head, and allow to cook for 10 minutes.

Remove and put in the already cut uziza and onions.

Taste for salt and pepper

Serve while it is still hot.

Bananas fritters.

Ingredients:

Bananas as required
Eggs, flour, nutmeg, olive oil, baking powder, salt pepper and brown sugar.

Wash, peel and mash banana in a bowl.

Add flour, salt and baking powder

Mix thoroughly.

Add milk, eggs, nutmeg and a little pepper and continue to mix until smooth.

Put oil in a frying pan and let it heat

With a table spoon, scoop the banana mixture and fry until brown or gold.

Banana fritters can be eating as a snack or dessert.

Nkwobi (Cow feet).

Ingredients:

Cow feet, onions, pepper, potash(Akuan), ugba, salt, palm oil, utazi, cray fish, maggi

Wash the cow feet, put in the pot, season with salt, maggi and onions and leave to cook until tender.

Add ground pepper and cray fish

Mix potash in a bowl of water, let it dissolve and then pour into the pot of cow feet.

Heat the palm oil and pour it in to the pot of the cow feet.

Stir until all the ingredients are absolve into the cow feet.

Allow to cook for 10 minutes. Taste for salt.

Cut the utazi and fresh onions thinly and sprinkle it on the nkwobi.

Good meal for the winter and raining season.

CHAPTER ELEVEN

JUICES

Orange Juice

Ingredients:

Several Oranges

Method:

Wash, Peel and slice the orange into two half and Blend with a juice blender.

Serve as part breakfast, or as desert.

Pineapple Juice

Ingredients:

One or two large pineapple.

Method:

Wash, peel and slice the pineapple into large pieces.

With a juice blender, blend the pineapple. Put in the refrigerator to cool and drink. It is best during the summer time as it help to cool off the heat. It could also be serve at any time of the day.

Pineapple and Carrot Juice

6 large carrots and 1 whole pineapple.

Wash carrot, slice into pieces and blend.

Wash pineapple remove skin slice into several pieces and blend together with the carrot.

Sieve and put in a container and place in the refrigerator to cool and then serve.

Fresh Tropical Blend

Ingredients:

2 mangoes, wash and peel, and take out the seeds.
3 oranges wash, peel and cut into two.
2 guavas wash and slice.
1 pawpaw, wash, slice and remove the seeds.
1 pineapple wash and remove skin and slice.

Method:

Put all the sliced fruits in a big bowl, and then blend all together

Put all the mix blended fruits together and store in a refrigerator to cool and then serve.

Tomatoes Juice

Ingredients:

Tomatoes, Cloves.
Honey, celery, fresh pepper.

Method:

Wash tomatoes, celery and fresh pepper slice and blend into a paste.

The celery and fresh pepper makes it spicy and tasty.

Strain with a strainer to get the clear juice out.

Wash and blend the clove and mix with tomatoes.

Turn it into a glass and add a little bit of honey to taste.

Put in a refrigerator to let it cool before drinking.

Afternoon delight

1 pineapple, top removed and skin.
2 large orange peeled.
6 strawberries.
2 packs red grape.

Method:

Wash pineapple and slice into cubes.

Wash and peel oranges and cut into two.

Wash strawberries and grapes.

Blend all together, store in a refrigerator and serve cool.

Orange and Pineapple Blast

2 large pineapple wash, remove skin and cut into cubes.

3 large oranges peel and cut into two.

Blend all together, serve when cool.

Ginger Drink

Ingredients:

Ginger, cloves, honey.

Method:

Wash and peel ginger. Cut it into tiny pieces and blend.

Add warm water and stir, and then sieve it.

Wash cloves, blend and add to the ginger drink to give it a beautiful flavor.

Add honey to taste and drink like tea either cold or hot.

It is best for the winter or cold weather, and one suffering from sinuses, cough or fever.

Oranges, pineapples, carrots can also be added to the ginger as a flavor.

Fresh Carrots, Apples and ginger Juice

Ingredients:

4 medium carrots.
2 large apples.
Ginger.

Method:

Wash, top and tail carrots.

Wash apple cut and remove stem.

Wash, peel ginger and blend the carrots, apple and ginger.

You can add more ginger if you want it spicy.

Sieve and keep in a refrigerator to cool and then serve.

Lemonade Juice

Ingredients:

6 large lemons.
Honey.
Water.

Method:

Wash and peel the lemon and blend the lemonade.

Add 2 cups of water and honey to taste.

Keep in a cool refrigerator and then serve.

Sugarless Lemonade

6 large lemons and 3 ripe apples.

Wash cut and remove stem. Wash and peel lemon.

Blend all together, sieve and store in a cool refrigerator and then serve.

Apple Juice

Ingredients:

Ten Apples.

Method:

Wash apples and take out stems. Cut the apples into smaller sections and then blend.

Mix sweet and tart apples for a delicious flavor.

Add carrot to taste and a vitamin booster.

Spice it with it up with sprig of mint or ginger.

Refrigerate the apple juice to let it cool and then drink.

Mango Juice

Ingredients:

Mangoes, cloves.

Wash, peel and slice mangoes to take out the seed and then blend.

Wash cloves, blend and add to the blended mango to give it a nice flavor

Refrigerate the mango juice to let it cool and then serve.

Carrot Juice

Ingredients:

Carrots, Ginger, honey and celery

Methods:

8 large carrots, wash, do not peel skins as it has vitamins and minerals, cut carrots and ginger into smaller pieces. Blend and sieve. Put in the refrigerator to cool and then serve.

Mango, kiwi and Carrot Juice

Ingredient:

3 large mangos.
4 kiwi fruits and 2 large carrots.

Method:

Wash mango, remove skin and pit.

Wash kiwi unpeel.

Wash carrot topped and tailed.

Put all in a blender and blend.

Leave to cool and then serve.

Fresh peach and Pear

Ingredients:

5 peaches and 2 pears

Method:

Wash peaches and pitted.

Wash pears peel and remove stem.

Blend all together. Keep to cool and the serve.

Orange and grapefruits Juice

Ingredients:

5 large oranges and 3 large grapefruits.

Method:

Wash and peel oranges and grapefruits.

Slice and blend.

Sieve and leave to cool and then serve.

Orange and grapefruits is use to boost vitamin C.

Fresh pear and Grape Juice

Ingredients:

5 pears and 2 bunches of seedless grapes.

Method:

Wash pears and remove stem.

Wash grapes and blend all together.

Leave to cool and serve.

Fresh Cucumber and carrot Juice

Ingredients:

6 large carrots.
Medium size cucumber.

Method:

Wash, topped and cut carrots into pieces.

Wash and slice cucumber. Blend carrots and cucumber. Put in a glass and store in the fridge to cool and then serve.

Cashew Juice

Ingredients:

20 cashew fruits, 2 lime, ginger and passion flavor syrup or honey and little water.

Method:

Blend cashew fruit, ginger and water until smooth. Strain and add lime juice. Add honey or passion flavored, leave to cool in a refrigerator and serve.

Guava Juice

Ingredients:

Guava, honey and warm water.

Method:

Peel the guava, take out the seeds and then blend.

With warm water and honey, mix the guava together.

Leave to cool and serve.

Strawberry Juice

Ingredients:

Fresh strawberry, honey and warm water

Method:

Wash and blend strawberries, add warm water. Blend until it is smooth.

Add sugar to taste. Keep in a refrigerator to cool and then serve.

Pawpaw Juice

Ingredients:

Pawpaw, water.

Peel pawpaw; remove the seeds from the core. Slice and then blend.

Store in the refrigerator to cool and then serve.

Soya Milk Drinks

Ingredients:

Soya beans, water.

Method:

Soak soya beans overnight, after removing the debris or the damaged soya beans. Boil the soya beans until soft and grind with a blender until it is slurry. Pour in a pot and add a little water and let it boil.

While boiling, keep stirring regularly or constantly and let it simmer for 15 to 25 minutes. Filter the slurry through a sieve or white linen cloth in a bowl. The left over pulp from the soya beans can be used for beans cake.

The soya beans or soy milk should be store in the refrigerator to cool before serving.

Zobo Drink

Ingredients:

Dried zobo leaves, ginger, flavors, garlic, lime or lemon juice, pineapple peels or juice.

Sally Iyobebe

Method:

Wash dried Zobo leaves properly in ordinary water to get rid of sand and particles. Put the Zobo in a cooking pot. Add moderate amount of water and allow to boil for about 25 minutes.

Add the ground ginger and garlic, the pineapple peels and the lemons rings and allow for 20 minutes.

Remove from heat, allow it to cool and sieve into a big bowl to remove residue. Add sugar or honey when still hot and stir properly to taste. Add flavors, like strawberry, pineapple, apple, orange and any other flavors used in baking. Put in bottles and put in a refrigerator immediately to prevent fermentation. Serve chilled. If too thick dilute with some water before serving.

Water Melon Juice

Ingredients

Water melon, knife and blender

Method

Wash and take the skin off the water melon

Put in a blender and blend until smooth

Sieve and put in a container.

Put in a refrigerator and leave to cool for 10 to 15 minutes

Serve cool and enjoy it.

Cantaloupe Juice

Cantaloupe is a refreshing water-based beverage. Made up of little more than fresh fruit, citrus juices, and water, are the perfect way to cool down when it's hot outside. Look for the ripest melons available in the produce section, the riper the melons the less sugar you'll have to add to the recipe. A ripe melon should be fragrant, heavy for its size, and should yield slightly to pressure at the stem end.

Ingredients:

1 or 2 large Cantaloupe, remove skin and cut into large dices.
4 cups of water.
3 teaspoons freshly squeezed lime juice.

Ice to cool the drink or put it in a refrigerator.

1. Place the cantaloupe, 2 cups of the water, the lime juice, in a blender and blend on high speed until smooth, about 20 seconds.
2. Strain through a fine-mesh strainer set over a large pitcher or bowl, using a ladle or wooden spoon to press down on the solids (you should have about 3 cups of liquid). Discard the solids.
3. Add the remaining 2 cups water and stir to combine. Taste and add additional sugar as needed. Refrigerate until cold, at least 1 hour. Serve over ice.
    cutting, cover your honeydew with plastic wrap or store in an air-tight container and use within a couple of days.

Honeydew Kiwi Cucumber Juice

This Honeydew Kiwi Cucumber Juice is a very different tasting juice, as it has a light stimulating taste. It is advisable not to use too much cucumber for it will outdo the flavor of the drink Use a minimal amount of cucumber to make the juice taste delicious.

Honeydew melon and kiwi fruit are superb sources of Vitamin C, which is an important vitamin for the repairs of healthy skin, hair and bones.

When you eat plenty of honeydew and kiwi which contains extraordinary ingredients of Vitamin C, you will be avoiding the risk of developing cardiovascular disease and other kinds of cancers like lung cancer, breast cancer and colon cancer.

It is good to drink your juice fresh after making it to avoid any harm or contaminations to your juice or store in the fridge for not more than a day.

Plantain Juice

Green plantain peels - Wash a green plantain and peel it, then put the peel in a jar and cover with water. Let it sit overnight, and then drink this water three times a day. Lowers your blood sugar level. Keep drinking as needed and change the peel every other day and refill the jar with water.

Hibiscus tea/Drink

Hibiscus tea is gotten from the hibiscus flower. It is similar to cranberry juice in taste. As a tea, hibiscus can be taken either hot or cold, depending on the person's preference. Drinking hibiscus tea regularly produces a range of health benefits to its user.

The Different Uses of Hibiscus

Because of the high medicinal nature of the Hibiscus plant, it has found many uses aside from being concocted into a zesty tea. In some countries, hibiscus is being used as a cooking ingredient. In many others, it is a popular first aid remedy for boils, bumps, cuts, and wounds. Hibiscus is used for lowering of high blood pressure, stroke and heart disease. It is also used to control high cholesterol levels, high inn Vitamin C and enhances the functions of liver.

There are many beneficial compounds present in hibiscus flower that can produce the choleretic effect, which is what allows the liver to function better.

Coconut Water Drink

This drink is naturally from fresh coconut. It is a clear liquid with fruit at the center of the palm.

It is best to drink the water when the coconut is still at the tender stage for you to enjoy it.

The coconut fruit, still very tender can be eating with a spoon

Best time to enjoy coconut water is during the summer time as it is very refreshing.

It is low in calories, Cholesterol free and contain more potassium than banana.

It helps in reducing cancer, flush out kidney stones and help in controlling hangover for those who drinks alcohol or are drunkards.

Aloe Vera Juice

Ingredients:

Aloe Vera plant, vinegar, water, orange juice knife and blender.

Method

Take off the leaf from the plant

With a sharp knife, peel the rind from the plant and take off the yellow layer in beneath

Put the rind inside the white vinegar cup of water

Continue the process until you are able to scoop about 2 or 3 tablespoon of clear Aloe Vera gel.

Sally Iyobebe

With a blender, and a cup of orange juice or mango juice, blende all together until it becomes smooth.

Pour in a glass or cup.

Store in a refrigerator and leave to cool for about 10 minutes

And then serve.

CHAPTER TWELVE

❦━━━━━━━━━━━━━━━❧

SMOOTHIES AND PUNCH DRINKS

Tropical Smoothie

Ingredients:

2 raspberries washed.
2 mangoes wash, peel and remove pit or stem.
1 pineapple wash, peel and remove top.
1 papaya (pawpaw) wash peel and remove seeds.
3 kiwis washed.

Method:

Juice the fruits. Mix all of them together and add ice to it.
Blend and thicken it to your taste.

Power punch

Ingredients:

3 cups parsley. Spinach, 2 medium apples, 2 stalks celery and 1 green or red bell pepper.

Wash parsley, spinach apple celery and bell pepper and blend together. Thicken it to your taste.

Sally Iyobebe

Skin Glow

Ingredients:

2 cucumber, I cup parsley 2 medium apples 5 carrots, coconut or soy milk.

Method:

Juice all the ingredients and add either coconut milk or soy milk.

Paradise Island

Ingredients:

1 Paw-paw (papaya,)2 oranges, 1 pineapple, lime and ginger.

Method

Wash, peel and remove seeds.

Wash and peel orange.

Wash, peel pineapple, remove top and cut into pieces.

Wash and peel lime and ginger, and slice.

Blend all together and thicken it to your taste.

Island Punch

Ingredients:

2 pineapples, 8 strawberries, 4 large apples and honey.

Method:

Wash, peel, remove top and cut into pieces.

Wash strawberries and remove stems.

Wash apples and remove stems.

Blend all together and add honey to taste.

Ginger Ale

Ingredients:

Ginger root, 1 cantaloupe, 6 strawberries and 2 oranges.

Method:

Wash, peel and slice ginger.

Wash, peel, slice and remove seeds from cantaloupe.

Wash, peel and slice the oranges.

Wash and remove stems from strawberries.

Juice all the ingredients together. Leave to cool and serve.

Sweet Dreams

Ingredients:

2 piece of ginger root.
2 stalks celery.
2 medium apples.
2 green peppers.
1 small cabbage.

Method:

Wash ginger, peel and slice.

Wash celery.

Wash green pepper.

Wash apples and remove stem.

Wash cabbage and slice it.

Bring all the ingredients into a juice form.

Blend all together and add 2 bananas and soy milk.

Breakfast pick-me up

Ingredients:

2 large carrots, 2 medium apples and 1 cantaloupe.

Method:

Wash; remove top and tail of carrot.

Wash apples and remove stem.

Wash cantaloupe, peel and remove seeds.

Blend all together. Leave to cool and then serve.

Beach body

Ingredients:

2 medium beets.
1 cup parleys or bundle of fresh parsley.
3 cucumbers.
2 celery fresh bundles.
6 carrots and 3 medium apples.

Method:

Wash apples, remove stem.

Wash celery, cucumber, beets and parsley.

Wash carrots; remove top and tail or bottom.

Blend all together. Leave to cool and then serve.

Passion Play

Ingredients:

Pomegranates.
Grapes.
Kiwis.
Strawberries.

Method:

Wash all the above fruits and blend. Leave to cool and serve.

Pear-Fect

Ingredients:

2 or 3 pears.
Fresh cranberries.
2 peaches.
1 pineapple.

Method:

Wash peaches and pit it.

Wash fresh cranberries.

Wash pears and removed stem.

Wash pineapple, peel skin and remove top and cut into pieces.

Blend all together. Leave to cool and then serve.

Raspberry Rush

Ingredients:

1 bundle raspberry.
2 large oranges.
1 small lime.

Method:

Wash and peel oranges and lime.

Wash raspberries.

Blend and serve with crushed ice or leave to cool and then serve.

Summer Punch

Ingredients:

1 Pineapple, strawberries, oranges and grapes.

Wash pineapple, remove skin, top and cut into pieces.

Wash strawberries and grapes.

Wash and peel oranges.

Blend all together and leave to cool and then serve.

Berry punch

Ingredients:

1 bundle cranberries, I bundle blue berries, 1 pack strawberries and raspberries.

Method:

Wash cranberries, blue berries, and strawberries

Blend all together and add a cup of sparklingly cider. Leave to cool and then serve.

Power up

Ingredients:

2 cucumber, fresh spinach, celery, 1 green pepper, and 2 large apples.

Sally Iyobebe

Method:

Wash celery, cucumber and spinach leaves and green bell pepper.

Wash 2 medium apples and remove stem.

Bring them into a juice form, store in a cool place and serve.

Kiwi Cooler

Ingredients:

4 Kiwi fruits.
1 pineapple.
3 oranges.
Ginger root.

Method:

Wash, peel and slice pineapple.

Wash and peel oranges.

Wash kiwis and wash and peel ginger.

Blend to a juice form and coconut milk and stir.

Keep in a refrigerator to cool and then serve.

Spicy orange

Ingredients:

1 large ginger.
4 large oranges.
1 small cantaloupe.
1 small lemon.

Method:

Wash and peel ginger.

Wash and peel oranges and lime.

Wash and peel cantaloupe.

Blend all together. Leave to cool ant then serve.

Salad in a glass

Ingredients:

2 fresh ripe tomatoes and 2 large carrots.
Lettuce, Celery, red bell pepper.

Method:

Wash tomatoes, lettuce, celery and red bell pepper.

Wash carrots, remove top and tail.

Blend in a juice form. Leave to cool and then serve.

Peachy keen

Ingredients:

2 peaches, 4 apricots, 3 pears and 2 medium apples.

Wash peaches and apricots.

Wash pears and remove stem.

Wash apples and remove stem.

Blend and leave to cool and then serve.

Strawberry special

Ingredients:

1 pack of strawberries.
1 lemon.
1/2 pack of raspberries.
Ginger ale.

Method:

Wash and remove stem from strawberries.

Wash raspberries.

Wash and peel lemon living the white pith.

Juice the raspberries, then the strawberries and then the lemon.

Stir the whole juice together in a bowl and then add the ginger ale.

Keep in the refrigerator to cool and then serve.

Fresh Strawberry and Pear Champagne

Ingredients:

Strawberries, pear and seltzer.

Method:

Wash strawberries.

Wash pears and remove stem.

Blend and add seltzer or white nonalcoholic champagne.

Leave to cool and then serve.

Love Potion

Ingredients:

1 pack of peaches.
1 pack of grapes.
1 pack of strawberries.
2 large apples.

Method:

Wash peaches, grapes and strawberries.

Wash apples and remove stems.

Blend and leave to cool and then serve.

Fresh Cran-Apple

1 pack of cranberries.
4 large apples.

Method:

Wash cranberries.

Wash apples and remove stems.

Blend and leave to cool and then serve.

Rainbow Surprise

Ingredients:

1 bundle of celery.
1 bundle of carrots.

2 cucumber, 2 yellow squash, 2 zucchini and 3 ripe sweet apples.

Method:

Wash celery, cucumbers, squash and zucchini.

Wash carrots, remove top and tail.

Wash apples and remove stem.

Blend all together and leave to cool and then serve.

Energy Boost

Ingredients:

Carrots, celery, beet root parsley lettuce, tomatoes spinach and watercress leave.

Method:

Wash carrots, remove top and tail.

Wash beets without taken the tops or roots off.

Wash and roll lettuce.

Wash and roll parsley, watercress and spinach.

Wash and stem tomatoes and add salt to taste.

Mix all together and leave to cool and then serve in a glass.

Cherry smoothie.

Ingredients:

Coconut milk, almond milk, raisings, cherries, oats and greens.

Organic spinach and lettuce.

Methods:

Wash organic spinach and lettuce and blend

Mix fresh coco milk, almond milk, cherries and raisins.

Add oats meal and mix all together.

Then serve cold.

Banana peach Kale.

Ingredients:

Lettuce or spinach

Banana, almond, peaches oatmeal

Wash lettuce and spinach and blend

Peel banana and peach and blend

Mix all the lettuce, spinach, banana, and peach.

Add the almond and oatmeal and mix all together.

Leave to cool and then serve.

Spinach smoothie.

Ingredients;

Spinach, mangoes, Pineapple, banana

Wash spinach, and slice it into pieces

Wash and Peel mangoes and slice and remove seed

Wash and peel pineapple and cut into pieces

Wash and peel banana

Put all in a blender and blend

Put in a bowl, leave to cool and serve.

CHAPTER THIRTEEN

VITAMINS AND SUPPLEMENTS

Vitamins.

Vitamins are said to be organic part of the food substance that are used for the maintenance of good health and life. Vitamins comes from the food we eat and since the body cannot produce enough nutrients, it is required that the body get a small amount of these vitamins to carry on the different purposes and functions the body needs to work properly. It is also needed for normal metabolism. Where vitamin is lacking in the food, it may result to deficiency disease.

There are two types of vitamins. Fat soluble and water soluble vitamins.

The vitamins can be broken down into different components.

Fat soluble vitamins are stored both in the fat tissues and liver in the body. Vitamins are easier to be produced in our body and stay for as a reserve for days and months. They are absolved in the intestinal tract with the assistant of fats. Fat soluble vitamins can be broken to down to Vitamin A, D, E, and K.

Water soluble vitamin do not have the capacity to store fat long enough in the body as they quickly do down through the urine. They need frequent replacement of the fat soluble vitamins. Water soluble vitamins contain vitamin C and B.

Types of vitamins.

Vitamin A contain carrots which is a food source of vitamins A and B3. And any deficiency of this vitamin may cause night- blindness and eye disorder. Food contents of this vitamin A are broccoli, carrots sweet potatoes, liver, eggs, apricot melon, milk cantaloupe cod liver oil and some cheese.

Vitamin B are water soluble and lack of it may cause beriberi. Food sources of vitamin B includes, yeast, brown rice, whole grain wheat, asparagus, liver, eggs, cauliflower, sweet potatoes, oranges cereals, grains and sunflower seeds.

B2 are water soluble and the food sources are asparagus, okra, meat, eggs, fish, milk yogurt, green beans banana etc.

B3 are water soluble and its food contents are chicken, beef, fish, milk, avocado, dates, sweet potatoes, tomatoes, carrots, nuts, whole grains, legumes, kidney, mushrooms, broccoli, and leafy vegetables. Deficiency may lead to pellagra.

B5 are water soluble, and contain broccoli, avocados, fish, whole grains and meat. Deficiency will lead to paresthesia.

B6 are water soluble and the food contents are bananas, nuts, meats and whole grains. Deficiency will cause anemia.

B7 are water soluble and the food contents includes egg yolk, vegetables and liver. Deficiency includes dermatitis and enteritis.

B9 are water soluble which includes leafy vegetables, sunflower seeds, legumes, liver and grains. Deficiency may cause still birth or birth defects in pregnant women.

B12 which is water soluble contains eggs, fish, meat. Cereals, poultry. Deficiency may lead to megaloblastic anemia.

Vitamin C another water soluble vitamins contains fruits and vegetables. Deficiency will cause megaloblastic anemia.

Vitamin D can be gotten from sun exposure and it contains, greens, fish, beef, liver, mushroom and eggs. Deficiency can cause rickets.

Vitamin E which is a fat soluble comprises of almond, nuts, avocado, eggs, wheat germ, kiwi fruit, milk and whole grains.

Vitamin K another fat soluble contains kiwi fruit, parsley, avocado and leafy green vegetables.

Supplements.

Supplements are substances that are gotten from the food we eat or drink. They can be in the form of minerals, herbs vitamins and plants which an individual use as protein used for the building block of the body. They can be in the form of tablets, capsule, pills, or liquid form. Good source of supplements is calcium, and vitamins D and E. Vitamin D is good for the maintenance of strong bones and for the reducing of bone loss. Folic acid is good to reduce the risk of birth defects. Omega-3 fatty acids from fish oils, helps in the cure of heart diseases.

Summary.

For weight lost to be achievable, it will be necessary to eat plenty of fruits and vegetables as they contain a lot of water. Eat plenty of it because of its high fiber contents, and are easier to digest than the solid food. Make sure you fruit or vegetables are in the liquid form for smooth metabolism. Processed foods, fried foods, eating of unhealthy foods like hamburger, pizza and fat will prevent easy digestion. Drink plenty of water to keep you hydrated, more energy, and proper functioning of the immune systems. Avoid the consumption of processed food, smoking, drinking of coffee and soda as it will cause the body to dehydrates.

What is our Relationship to Food?

That food is a basic and essential to the extent that we cannot get away from it. For if we have to survive and thrive, we must have to eat. But when we over indulged in it, it becomes too bad and gluttony. We love to eat because it nurtures our body and therefore replaces our sense of wellbeing. But while craving for that love, we turn to food and we eat more of it turning it to our friend, lover, partner and therapist.

Again food is a beautiful part of our lives. As a caterer and former owner of a restaurant, I will like to invite people to gracefully make ways to turn food into a unique ceremony that benefits our lives on every aspect. Rather than making food a person, let us make food a way of beautification. It should be that which expresses our delicacies as a woman and as a human being.

When food is overindulged in, it takes on the qualities of the food becomes the master and the eater becomes the servant. So you are constantly in need of pleasing the food and therefore a split between you and the food.

We should also focus where our food comes from, for real food is like real love. Which is born of the earth, grows like some kind of miracle that we take for granted. We should remember where our food has come from, the journey from field to plate.

This book is unique in that, it has a lot of recipes mainly African, and first of its kind. To enable you try your hands on some of the recipes, I will be glad enough to assists you in where to get the stuff, and how to prepare it. We can also prepare the food as a take away for a fee. We cannot keep pretending to think all is well when in the actual sense it is not. But because we have enslaved ourselves in eating junk food, and going the easy way, and being lazy to cook, and the fact that we do not want to clean, we have prevented ourselves from eating right. Therefore, we find ourselves paying more on medication and hospital bills, money which we could safe if only we did it right. One thing I gain from late Mrs. Stella Obasanjo, former first lady of Nigeria, while in Aso Rock, she never let anybody prepared her husband's, the then president' food. It was really impressive. A very busy lady who still found time in preparing her husband's food. She had all the money to get best cooks but she decided against it. I used to weigh 220 Ibs. In 6 months I cut down to 150 lbs. and I felt really good. My mom; daughter, families, and friends who came to welcome me last year at the Muritala Mohammed Airport in Lagos could not even recognized me. But one thing I felt good inside me and became a role model to others. So if you are ready in this year 2016 to cut down on your weight, I will be ready to help coach you in whatever way that I can. You do not have to be ashamed to approach me. Let us work together, eat right, stay healthy, and spend less on hospital bills.

Printed in the United States
By Bookmasters